BASIC PRINCIPLES

OF

ENGINEERING

BASIC PRINCIPLES

OF

ENGINEERING

First edition (2017)

Arjun Ghimire

To

My beloved family

Preface

This textbook provides both profound technological knowledge and a comprehensive treatment of essential topics in basic engineering. Including numerous examples, figures and exercises, this book on **"Basic Principles of Engineering"** is suited for students, lecturers and researchers working in the general field of engineering of all disciplines.

We are indebted to several authors, whose published materials we took the liberty to use. As a token of thanks, we have appended their work in the bibliography.

Data, including properties and charts, are provided, but for definitive design details may need to be independently checked to ensure requisite precision. Every effort has been made to provide clear explanations and to avoid errors, but errors may occur though.

So feedback from users will be most welcome, and should be directed to the author.

Arjun Ghimire
(Lecturer)
Central Campus of Technology
Tribhuvan University
Dharan, Nepal

Contents

List of figures

List of tables

List of solved numericals

List of problems

List of equations

Chapter 1: Units, Dimensions and Conversions

Physical quantities

The quantities by means of which we describe the laws of physics are called physical quantities. A physical quantity is a quantity in physics that can be measured or that can be quantified. Examples of physical quantities are mass, amount of substance, length, time, temperature, electric current, light intensity, force, velocity, density, and many others.

There are two type of physical quantities.

i. Fundamental quantities

ii. Derived quantities

Fundamental quantities

Physical quantities which are independent of each other and cannot be further resolved into any other physical quantity are known as fundamental quantities.

Table 1.1: Seven fundamental quantities

Fundamental Quantity	Unit	Symbol
Length (L)	Metre	m
Mass (m)	Kilogram	Kg
Time (t)-	Second	Sec/ s
Amount of substance	Mole	Mol
Temperature	Kelvin	K
Electric current	Ampere	A
Luminous intensity	Candela	Cd

Derived Quantities

Physical quantities which depend upon fundamental quantities or which can be derived from fundamental quantities are known as derived quantities.

Table 1.2: Some examples of derived units

Derived Quantity	Unit	Symbol
Force	Newton	N
Energy	Joule	J
Power	Watt	W
Pressure	Pascal	Pa
Frequency	Hertz	Hz
Electric charge	Coulomb	C
Electrical potential	Volt	V
Electrical resistance	Ohm	Ω

Units

Things in which quantity is measured are known as units.
Measurement of physical quantity = (Magnitude) × (Unit)
All physical quantities consists of 2 parts:

i. *Unit:* It indicates about the quantity and gives standard by which it is measured. Eg. cm, foot, sec, etc.

ii. *Number:* It denotes the number of units needed. Eg. One, two, three, etc.

There are three types of units:

i. Fundamental or basic units

ii. Derived units

iii. Supplementary units

Fundamental or basic units

Units of fundamental quantities are called fundamental units. E.g. meter, kilogram, second, etc.

Characteristics of fundamental units

i. They are well defined and are of a suitable size.

2

ii. They are easily reproducible at all places.

iii. They do not vary with temperature, time, and pressure etc. i.e. invariable.

iv. There are seven fundamental units.

Derived units

Secondary units or derived units are those that are made by the inclusion of primary units. Units of derived quantities are called derived units.

Eg: Volume = (length)^3 (m^3); Speed = length/time (m/s)

Supplementary units

The units defined for the supplementary quantities namely plane angle and solid angle are called the supplementary units. The unit for plane angle is rad and the unit for the solid angle is steradian.

The supplemental quantities have only units but no dimensions.

Principle system of units

All physical quantities, met within this subject, are expressed in terms of the following three fundamental quantities:

a. Length (*L* or *l*), **b.** Mass (*M* or *m*), and **c.** Time (*t*).

For the purpose of measurement (dimensions), 3 systems have been used namely:

i. CGS system (Centimeter- Gram- Second) – Also called Metric system

ii. FPS system (Foot- Pound- Second) – Also called British system.

iii. MKS or SI system (Meter- Kilogram- Second) – Also called Modern system. In 1971, the international Bureau of weight and measures held its meeting and decided a system of units which is known as the international system of units.

Dimensions and dimensional formula

Dimensions of a physical quantity are the powers to which the fundamental quantities must be raised to represent the given physical quantity. Dimensional formula is an expression which shows how and which of the fundamental units are required to represent the unit of physical quantity.

E.g. Force (Quantity) = mass × acceleration

$$= mass \times \frac{Velocity}{(time)}$$

$$= mass \times \frac{length}{(time)^2} = mass \times length \times (time)^{-2}$$

So dimensions of force: 1 in mass, 1 in length and –2 in time.

And, Dimensional formula = $[MLT^{-2}]$

Table 1.3: Dimensional formula of different quantities

Quantity	Symbol	Formula	S.I. Unit	D.F
Displacement	S	-	Metre or m	M^0LT^0
Area	A	l ×b	(Metre)^2 or m^2	$M^0L^2T^0$
Force	F	m × a	Newton or N	MLT^{-2}
Work	W	F ×d	N.m	ML^2T^{-2}
Power	P	W/ t	Watt or W	ML^2T^{-3}
Impulse	-	F ×t	N.sec	MLT^{-1}

Inter- conversions

Any quantity can be converted from one system to other by the use of conversion factors. Conversion factor is a pure number and simply the ratio of the magnitude of the unit in one system to that of the corresponding unit in the other system.

Conversion factors are simply the multiplication factors. Eg. The length expressed in FPS system is 10 ft. To convert it from FPS to SI system, the factor is 0.3048 i.e. 1 foot = 0.3048 metre.

In the above example, the conversion factor is 0.3048. Metre (MKS) has higher magnitude compared to feet (FPS) system. Therefore, metre has a small number compared to the corresponding number in feet. Hence, foot has to be multiplied by the conversion factor to get metre.

Therefore, 10 feet = 10 × Conversion factor = 10 × 0.3048 = 3.048m. Conversely, metre should be divided by conversion factor to get feet i.e. 3.048 metre = $\dfrac{1}{0.3048}$ ×3.048 = 10 ft.

Table 1.4: Conversion table

1 ft = 12 in = 0.3048 m	1 BTU = 1055.06 J = 0.25216 Kcal
1 in = 2.54 cm	°F = 32 + 1.8 × °C
1 US gallon = 3.7854 L	$°C = \dfrac{\left(°F - 32\right)}{1.8}$
1 KWh = 3600 KJ	
1 lb$_m$ = 0.4536 Kg	°R = 460 + °F
1 lb$_f$ = 4.4482 N	°K = 273.15 + °C
1 psi = 6894.76 Pa	$\Delta\,°F = \dfrac{\Delta°C}{1.8}$
°F = 32 + 1.8 × °C	
1 HP = 745.7 W	$\Delta\,°C = \Delta\,K$; $\Delta°F = \Delta\,°R$

Example 1.1: Conversion of g/ml to Kg/m³

1. The density of talc is reported as 2.7 g/ml. Express the same in SI system (Kg/m³).

Sol^n: $\dfrac{g}{ml}$ is equal to $\dfrac{g}{cm^3}$

$1\ g = \dfrac{1}{1000}\ kg;\ 1\ cm^3 = \dfrac{1}{(100)^3}\ m^3.$

Therefore, $\dfrac{g}{cm^3} = 1000 kg/m^3$ and 1000 is the *conversion factor*.

Hence, 2.7 g/ml = 2.7 × 1000 = 2700 Kg/m³.

Exercises on unit conversions

1. Make the following conversions:

i. 251°F to °C (**Ans:** 121.7 °C)

ii. 500°R to K (**Ans:** 277.6 K)

iii. 0.04 lb_m/in^3 to kg/m³ (**Ans:** 1107.2 kg/m³)

iv. 12000 Btu/h to W (**Ans:** 3516.9 W)

v. 32.174 ft/s² to m/s² (**Ans:** 9.807 m/s²)

vi. 0.01 ft²/h to m²/s (**Ans:** 2.58x10^{-7} m²/s)

vii. 0.8 cal/g°C to J/kgK (**Ans:** 3347.3 J/kgK)

viii. 20000 kg m/s² m² to psi (**Ans:** 2.9 psi)

ix. 0.3 Btu/lbm°F to J/kgK (**Ans:** 1256 J/kgK)

x.1000ft³/(hft²psi/ft) to cm³/(scm²Pa/cm) (**Ans:**0.0374cm³/(scm²Pa/cm)

xi. 100 Btu/hft²°F to KW/m²°C. [**Ans:** 0.5678 KW/ m²°C]

xii. 100lbmol/hft² to Kgmol/sm². [**Ans:** 0.1356 Kgmol/sm²]

xiii. 0.5 lbf.s/ ft² to Pa.s.[**Ans:** 23.94 Pas]

2. In the literature, mass transfer coefficients in the gas phase are often reported in terms of lb_{mol}/hft^2atm. Determine the conversion factor by which the above must be multiplied in order to obtain the corresponding value of $kgmol/sm^2Pa$.[**Ans:** 1.337×10^{-8}]

3. The overall coefficient of heat transfer is 200 Btu/hft^2F. Convert the same into SI units (W/m^2K). [**Ans:** 1135.74 W/m^2K]

4. The unit of viscosity in the CGS system is Poise which is equal to $1dyne.s/cm^2$. If the fluid has a viscosity of 0.20 poise, calculate the corresponding value in SI units (Pa.s). [**Ans:** 0.02 Pas]

5. The viscosity of water at 60°F is given as 7.8 x 10^{-4} lb ft^{-1} s^{-1}. Calculate this viscosity in N s m^{-2}. [**Ans:** 1.16 x 10^{-3} N s m^{-2}]

6. The thermal conductivity of aluminium is given as 120 Btu ft^{-1} h^{-1} °F-1. Calculate this thermal conductivity in J m^{-1} s^{-1} °C^{-1}. [**Ans:** 208 J m^{-1} s^{-1} °C^{-1}]

Chapter 2: Refrigeration

Introduction

Refrigeration is the cooling effect of the process of extracting heat from a lower temperature heat source, a substance or cooling medium, and transferring it to a higher temperature heat sink, probably atmospheric air and surface water, to maintain the temperature of the heat source below that of the surroundings.

Refrigeration is a process in which work is done to move heat from one location to another. The work of heat transport is traditionally driven to mechanical work, but can also be driven by magnetism, laser or other means. Refrigeration has many applications, including, but not limited to; household refrigerators, industrial freezers, cryogenics, air- conditioning and heat pumps.

Primary purpose of refrigeration

Rates of decay and of deterioration in foodstuffs depend on temperature. At suitable low temperatures, changes in the food can be reduced to economically acceptable levels. The growth and metabolism of micro-organisms is slowed down and if the temperature is low enough, growth of all microorganisms virtually ceases.

Enzyme activity and chemical reaction rates (of fat oxidation, for example) are also very much reduced at these temperatures. To reach temperatures low enough for deterioration virtually to cease, most of the water in the food must be frozen.

It is well-known that some bacteria are responsible for degradation of food, and enzymatic processing cause ripening of the fruits and vegetables. The growth of bacteria and the rate of enzymatic processes are reduced at low temperature. This helps in reducing the spoilage and improving the shelf life of the food. Table below shows useful storage life of some plant and animal tissues at various temperatures. It can be seen that the storage temperature affects the useful storage life significantly. In general the storage life of most of the food products depends upon water activity, which essentially depends upon the presence of water in liquid form in the food product and its temperature. Hence, it is possible to preserve various food products for much longer periods under frozen conditions.

Table 2.1: Effect of temperature on storage life

Food Product	Average useful storage life (days)		
	0°C	22°C	38°C
Meat	6-10	1	<1
Fish	2-7	1	<1
Poultry	5-18	1	<1
Dry meats and fish	>1000	>350& <1000	>100 & <350
Fruits	2-180	1-20	1-7
Dry fruits	>1000	>350 & <1000	>100 & <350
Leafy vegetables	3-20	1-7	1-3
Root crops	90-300	7-50	2-20
Dry seeds	>1000	>350 & <1000	>100 & <350

Heats in Refrigeration

Refrigeration is simply concerned with the two heats: Sensible and Latent.

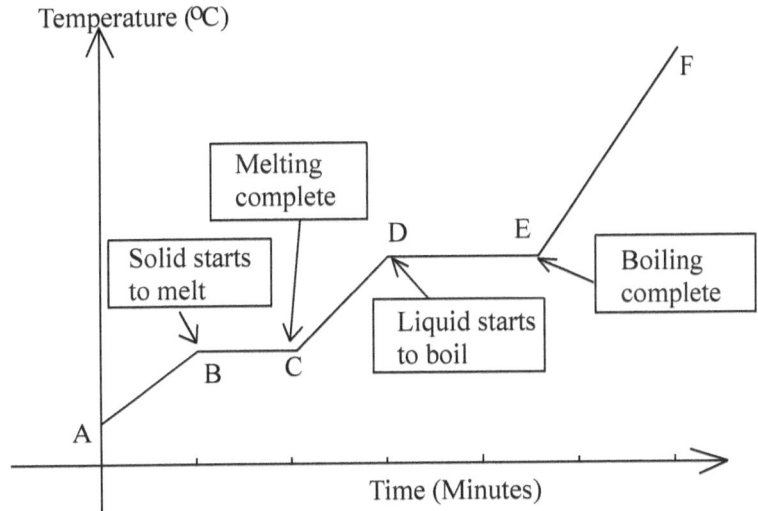

Fig. 2.1: Heats in refrigeration

For any substance, there are 2 distinct quantities of heat involved in the change of state.

Latent heat of fusion

It is the quantity of heat that must be absorbed by the substance in the solid state in order to be changed to a liquid or that must be removed from liquid to change to solid state. It is shown in the Fig. 2.1 from B to C.

Latent heat of vaporization

It is the quantity of heat which must be absorbed by the substance in the liquid state in order to change to a gas or that must be removed from gas if it is to return to a liquid state. It is shown in the Fig. 2.1 from D to E.

11

Principles of Refrigeration

There are 2 principles of refrigeration system:

i. Compression system

ii. Absorption system

The basic difference between them lies in the source of energy that drives them. The compression system relies on mechanical energy produced by a motor driven compressor. The absorption system uses heat energy supplied by a flame or some other sources of heat.

Vapour Compression Refrigeration Cycle [VCRC]

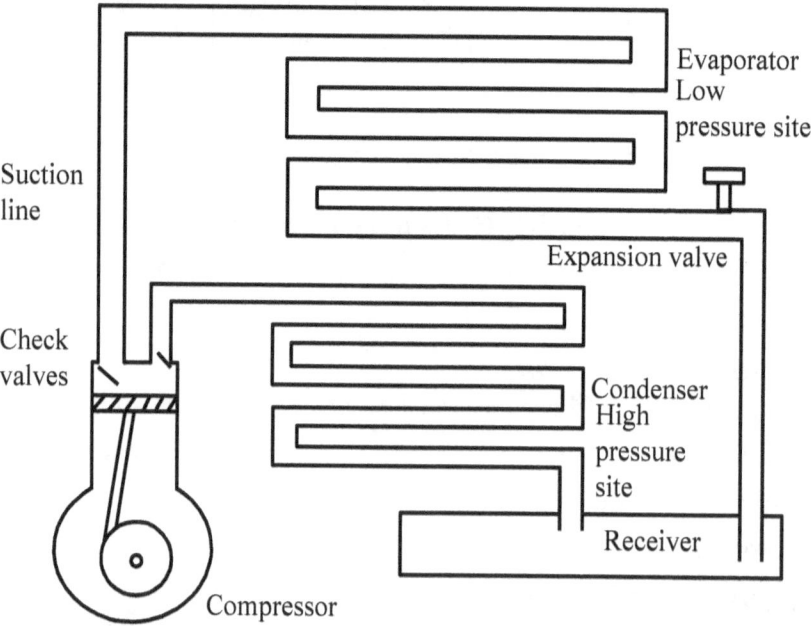

Fig. 2.2: Vapour compression refrigeration cycle

A compressor is generally a cylinder in which a piston works. The high pressure is changed to low pressure liquid so that it can evaporate at low temperature by means of expansion valve. The cycle is thus completed.

In these systems, a compressor compresses the refrigerant to a higher pressure and temperature from an evaporated vapour at low pressure and temperature. The compressed refrigerant is condensed into liquid form by releasing the latent heat of condensation to the condensed water. Liquid refrigerant is then throttled to a low pressure, low temperature vapour producing the refrigeration effect during evaporation. Vapour compression is often called mechanical refrigeration i.e. refrigeration by mechanical compression.

Components of VCRC

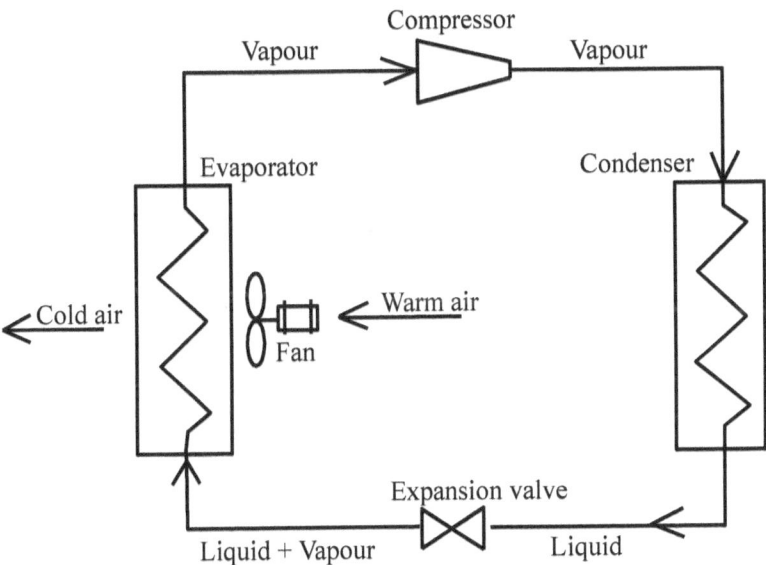

Fig.2.3: Components of VCRC

It consists of the following essential parts:

i. Compressor: The low pressure and temperature vapour refrigerant from evaporator is drawn into the compressor through the inlet or suction valve, where it is compressed to a high pressure

13

and temperature. This high pressure and temperature vapour refrigerant is discharged into the condenser through the delivery or discharge valve.

ii. Condenser: The condenser or cooler consists of coils of pipe in which the high pressure and temperature vapour refrigerant is cooled and condensed.

The refrigerant, while passing through the condenser, gives up its latent heat to the surrounding condensing medium which is normally air or water.

iii. Receiver: The condensed liquid refrigerant from the condenser is stored in a vessel known as receiver from where it is supplied to the evaporator through the expansion valve or refrigerant control valve.

iv. Expansion Valve: It is also called throttle valve or refrigerant control valve. The function of the expansion valve is to allow the liquid refrigerant under high pressure and temperature to pass at a controlled rate after reducing its pressure and temperature. Some of the liquid refrigerant evaporates as it passes through the expansion valve, but the greater portion is vaporized in the evaporator at the low pressure and temperature.

v. Evaporator: An evaporator consists of coils of pipe in which the liquid-vapour refrigerant at low pressure and temperature is evaporated and changed into vapour refrigerant at low pressure and temperature. In evaporating, the liquid vapour refrigerant absorbs its latent heat of vaporization from the medium (air, water or brine) which is to be cooled.

Pressure enthalpy chart of refrigerant

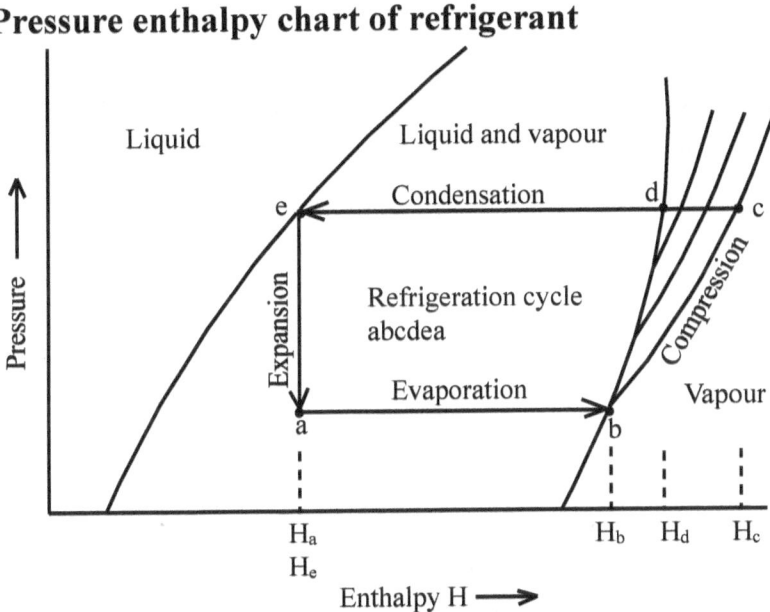

Fig. 2.4: Pressure-enthalpy chart of refrigerant

To start with the evaporator; in this the pressure above the refrigerant is low enough so that evaporation of the refrigerant liquid to a gas occurs at some suitable low temperature determined by the requirements of the product.

On the pressure/enthalpy chart this is represented by ab at constant pressure (the evaporation pressure) in which 1 kg of refrigerant takes in (H_b - H_a) kJ . The low pressure necessary for the evaporation at the required temperature is maintained by the suction of the compressor.

The remainder of the process cycle is included merely so that the refrigerant may be returned to the evaporator to continue the cycle. First, the vapour is sucked into a compressor which is essentially a gas pump and which increases its pressure to exhaust it at the higher pressure to the condensers.

This is represented by the line bc which follows an adiabatic compression line, a line of constant entropy (the reasons for this must be sought in a book on refrigeration) and work equivalent to $(H_c - H_b)$ kJ kg^{-1} has to be performed on the refrigerant to effect the compression.

To complete the cycle, the refrigerant must be condensed, giving up its latent heat of vapourization to some cooling medium. This is carried out in a condenser, which is a heat exchanger cooled generally by water or air.

Condensation is shown on figure along the horizontal line (at the constant condenser pressure), at first cd cooling the gas and then continuing along de until the refrigerant is completely condensed at point e. The total heat given out in this from refrigerant to condenser water is $(H_c - H_e) = (H_c - H_a)$ kJ kg^{-1}.

Expansion through the expansion valve nozzle is at constant enthalpy and so it follows the vertical line ea with no enthalpy added to or subtracted from the refrigerant. This line at constant enthalpy from point e explains why the point a is where it is on the pressure line, corresponding to the evaporation (suction) pressure.

Overall, the energy side of the refrigeration cycle can therefore be summed up:

a. heat taken in from surroundings at the (low) evaporator temperature and pressure $(H_b - H_a)$,

b. heat equivalent to the work done by the compressor $(H_c - H_b)$

c. heat rejected at the (high) compressor pressure and temperature $(H_c - H_e)$.

16

A useful measure is the ratio of the heat taken in at the evaporator (the useful refrigeration), (H_b - H_a), to the energy put in by the compressor which must be paid for (H_c - H_b). This ratio is called the coefficient of performance (COP).

Refrigerants

Refrigerant is a fluid used for heat transfer in a refrigerating system that absorbs heat during evaporation from a region of low temperature and pressure, and releases heat during condensation at a region of higher temperature and pressure. Fluorocarbons, especially chlorofluorocarbons, became commonplace in the 20th century, but they are being phased out because of their ozone depletion effects. Other common refrigerants used in various applications are ammonia, sulfur dioxide, and non-halogenated hydrocarbons such as propane

Types of refrigerants

According to the mode of working, refrigerants can be divided into two types:

Primary refrigerants

Primary refrigerants are those that pass through the process of compressor, condenser, expansion and evaporation during cyclic processes. Eg. Ammonia, R- 12, R- 22, CO_2, etc.

Secondary refrigerants

The medium which does not go through the cyclic processes in a refrigeration system and is only used as a medium for heat transfer are referred to as secondary refrigerants. Water, brine solutions of sodium chloride and calcium chloride come under this category.

Properties of ideal refrigerants

The properties of ideal refrigerants are enlisted as follows:

i. The refrigerant should have low boiling point and low freezing point.

ii. It must have low specific heat and high latent heat. Because high specific heat decreases the refrigerating effect per kg of refrigerant and high latent heat at low temperature increases the refrigerating effect per kg of refrigerant.

iii. The pressures required to be maintained in the evaporator and condenser should be low enough to reduce the material cost and must be positive to avoid leakage of air into the system.

iv. It must have high critical pressure and temperature to avoid large power requirements.

v. It should have low specific volume to reduce the size of the compressor.

vi. It must have high thermal conductivity to reduce the area of heat transfer in evaporator and condenser.

vii. It should be non-flammable, non-explosive, non-toxic and non-corrosive.

viii. It should not have any bad effects on the stored material or food, when any leak develops in the system.

ix. It must have high miscibility with lubricating oil and it should not have reacting properly with lubricating oil in the temperature range of the system.

x. It should give high COP in the working temperature range. This is necessary to reduce the running cost of the system.

xi. It must be readily available and cheap.

18

Units of refrigeration

The practical unit of refrigeration is expressed in terms of "tonne of refrigeration". A tonne of refrigeration is defined as the amount of refrigeration effect produced by the uniform melting of one tonne (1000 Kg) of ice from and at 0°C in 24 hours.

Since the latent heat of ice is 335 KJ/ Kg, therefore 1 tonne of refrigeration:

1 TR = 1000 × 335 KJ in 24 hrs

$$= \frac{1000 \times 335}{(24 \times 60)} = 232.6 \text{ KJ/ min} \dots\dots\dots \text{ (2.1)}$$

Naming of halocarbon compounds

They are represented by a 3 digit nomenclature. The first digit represents the number of carbon atoms in the compound minus one, the second digit stands for the number of hydrogen atom plus one while the third digit stands for the number of fluorine atoms. The remaining atoms are chlorine.

e.g., **R-22**
Number of C-atoms: C-1 =0, Or, C =1;
Number of H- atoms: H+1 = 2, Or, H = 1;
Number of F- atoms: F=2
Cl = 1 as C is tetravalent.

$$
\begin{array}{c}
\text{Cl} \\
| \\
\text{F}-\!\!-\text{C}-\!\!-\text{F} \\
| \\
\text{H}
\end{array}
$$

Hence, R-22 is $CHClF_2$ [Monochloro- difluoro- methane]

R-134

Number of C- atoms: C-1= 1, Or, C = 2;
Number of H-atoms: H+1 = 3, Or, H = 2;
Number of F- atoms: F =4
Number of Cl atoms =0.

19

Hence, R-134 is $C_2H_2F_4$ [Tetrafluoroethane]
Similarly, **R-40** is Methyl chloride.

Some refrigerants and their properties
Dichloro-difluoro methane (CCl_2F_2)

It is commercially known as Freon-12 or R-12. It is a clear colourless liquid which boils at 22.8°C at atm. Pressure. It is non-toxic and non- flammable and hence preferred for wide varieties of applications. It is miscible with lubricants and good electric insulator. These properties make it a good choice for use in sealed type domestic refrigerators.

F-22

Freon 22 has boiling point of -40°C at atm pressure. It is equally miscible with oil. As compared to F-12, its latent heat is lower and discharge temperature at compressor is higher and requires cooling of compressor though it requires smaller compressor displacement.

F-11

F-11 is used in chilling and air- conditioning purpose. At low pressure, high volume of this refrigerant is required for producing desired refrigerating effect and for this reason, centrifugal compressor is used in conjunction with F-11.

It has now been found that the refrigerants of Freon series causes depletion of ozone layer which is important for obstructing UV radiation from the sun. Phasing out of these refrigerants is a major problem which the world is attempting to handle.

Inorganic compounds

These compounds include ammonia (NH_3), water (H_2O), and gases used in the gas expansion systems. As refrigerants, they were used far earlier than the halocarbons. Air is a mixture of nitrogen, oxygen, argon, rare gases, and water vapour. Air has zero ozone depletion and is a zeotropic blend that has a temperature glide of 5.5°F (3°C) at atmospheric pressure. Ammonia also has zero ozone depletion. It has a high operating pressure at 40°F (4.4°C) evaporating and 100°F (37.8°C) condensing. Ammonia compressors show a smaller cfm/ton displacement and higher energy efficiency than HCFC-22 compressors. Leakage of ammonia is easily detected due to its objectionable odour. Ammonia attacks copper even in the presence of a small amount of moisture. It is of higher toxicity. An ammonia-air mixture is flammable if the concentration of NH_3 by volume is within 16 to 25 percent. The mixture may explode if the ignition source is above 1200°F (650°C). Because the safety classification of ammonia is B2 i.e. lower flammability and higher toxicity, it is not allowed to be used in comfort air conditioning in the United States.

Water has a zero ozone depletion potential and is readily available. At 40°F (4.4°C) evaporating and 100°F (37.8°C) condensing, water's evaporating and condensing pressures are both below atmospheric. Air and other non-condensable gases must be purged out of the refrigeration system periodically.

Chapter 3: Boilers

Steam

Steam is a state of water, partially or fully vaporized. It is the technical term for water vapour, the gaseous phase of water, which is formed when water boils.

Types of steam

Wet steam

Water vapours that includes water droplets is known as wet steam. As wet steam is further heated, the water droplets evaporate and at high enough temperature, all of the water evaporates and the system is in vapour- liquid equilibrium.

The wet steam is one containing water particles and is not treated as perfect gas but once it is saturated it can be regarded as perfect and in the region of superheat, it is treated as perfect gas. The quality of wet steam is described by:

Dryness factor, 'q':

$$q = \frac{M_s}{M_s + W} \ldots\ldots\ldots (3.1),$$

where M_s is the mass of saturated steam in total steam mass of $(M_s + W)$, W being the mass of water particles.

Superheated steam

It is steam at a temperature higher than its boiling point for the pressure which only occurs when all the water has evaporated or has been removed from the system.

__Advantages of superheated steam__

1. It has the ability to carry large amount of heat.
2. Cheap and readily available.

3. Suitability for heating after its use for power production.

4. Harmless being non- toxic.

5. High temperature gives higher thermal efficiency.

6. It can be expanded considerably before its temperature falls so low so as to cause condensation. So, heat lost due to condensation is reduced.

7. As superheated steam has high heat content, the specific steam consumption is low, which results in small sized plants.

Boilers

These are the steam generating devices. A basic diagram of a boiler is shown in Fig. 3.1. This diagram shows that a boiler comprises two separate systems. One system is the steam-water system, which is also called the water side of the boiler. Into this system water is introduced and, upon receiving heat that is transferred through a solid metal barrier, is heated, converted to steam, and leaves the system in the form of steam.

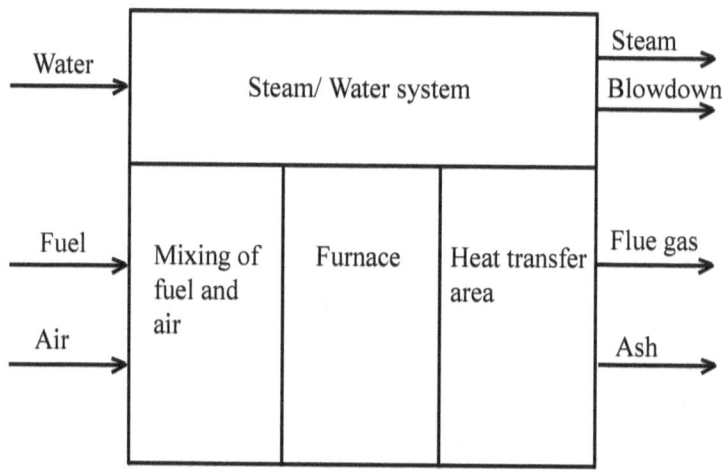

Fig. 3.1: Basic diagram of a boiler

The other system of a boiler is the fuel-air-flue gas system, which is also called the fire side of the boiler. This system provides the heat that is transferred to the water. The inputs to this system are fuel and the necessary air required to burn the fuel.

In this system the fuel and air are thoroughly mixed and ignited in a furnace. The resulting combustion converts the chemical energy of the fuel to thermal or heat energy. The furnace is usually lined with heat transfer surface in the form of water-steam circulating tubes. These tubes receive heat radiating from the flame and transfer it to the water-side system. The gases resulting from the combustion, known as the flue gases, are cooled by the transfer of their heat by what is known as the radiant heat transfer surface. The gases leave the furnace and pass through additional heating surface that is in the form of water-steam circulating tubes. In this area the surfaces cannot "see" the flame, and the heat is transferred by convection. Also in this area, known as the convection heating surface, additional amounts of heat are transferred to the water side of the boiler. This heat transfer further cools the flue gases, which then leave the boiler.

Types of boilers

According to the relative position of water and hot gases:

i. Fire tube or smoke tube boiler

ii. Water tube boiler

Fire- tube or smoke- tube boilers

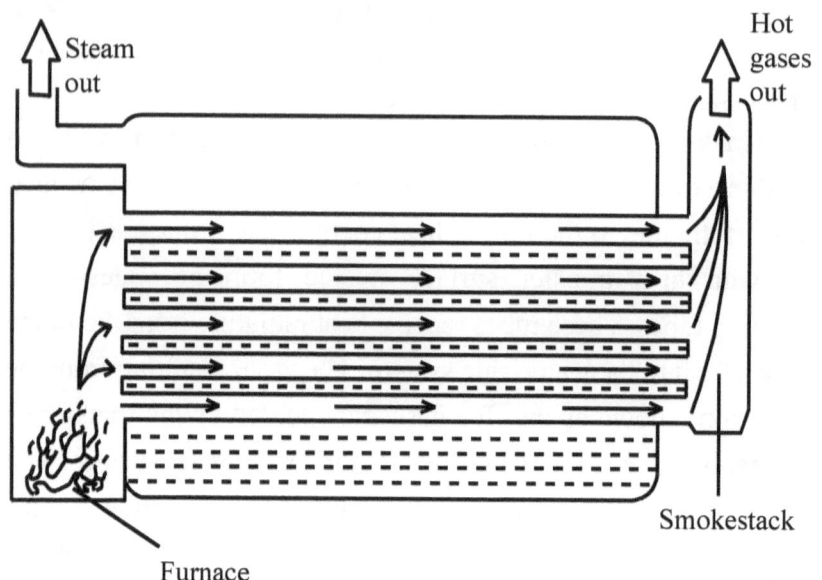

Fig. 3.2: Fire tube boiler

Fire tube or "fire in tube"boilers contain long steel tubes through which the hot gases from a furnace pass and around which the water to be converted to steam circulates. Fire- tube boilers typically have a lower initial cost, more fuel efficient and easier to operate, but they are limited generally to capacities of 25 tons/hr and pressures of 17.5 Kg/m^2.

Firetube boilers constitute the largest share of small- to medium-sized industrial units. In firetube boilers the flue gas products of combustion flow through boiler tubes surrounded by water. Steam is generated by the heat transferred through the walls of the tubes to the surrounding water. The flue gases are cooled as they flow through the tubes, transferring their heat to the water; therefore, the cooler the flue gas, the greater the amount of heat

26

transferred. Cooling of the flue gas is a function of the heat conductivity of the tube and its surfaces, the temperature difference between the flue gases and the water in the boiler, the heat transfer area, the time of contact between the flue gases and the boiler tube surface, and other factors.

Firetube boilers used today evolved from the earliest designs of a spherical or cylindrical pressure vessel mounted over the fire with flame and hot gases around the boiler shell. This obsolete approach has been improved by installing longitudinal tubes in the pressure vessel and passing flue gases through the tubes. This increases the heat transfer area and improves the heat transfer coefficient.

Lancashire boiler

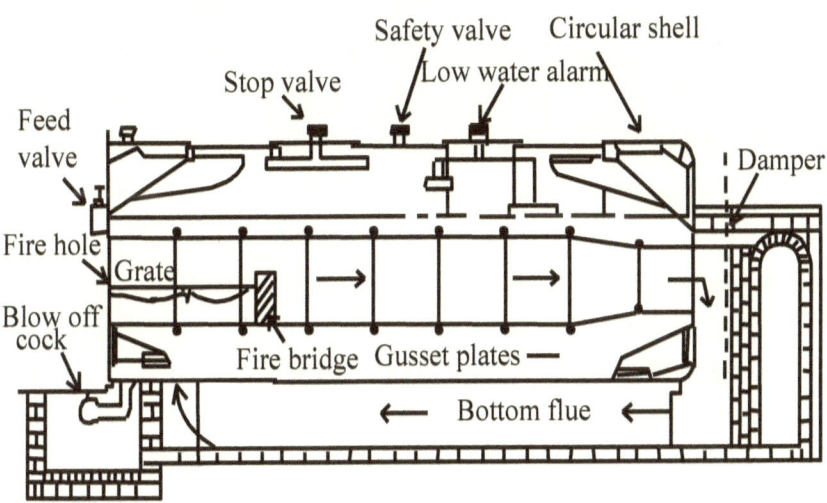

Fig. 3.3: Front view of Lancashire boiler

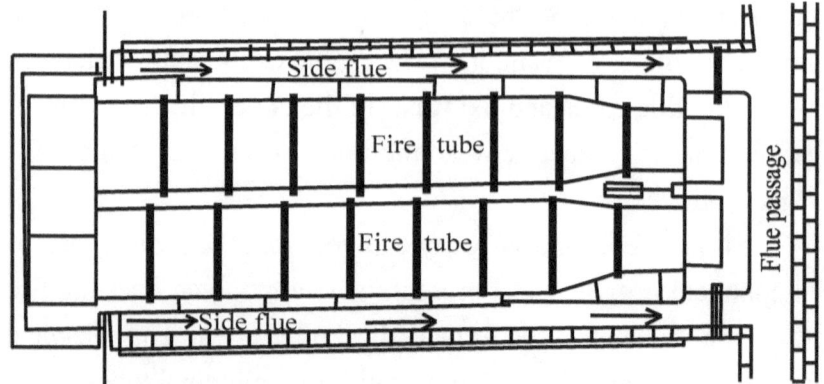

Fig. 3.4: Top view of Lancashire boiler

Coal is fed to the grate through the fire hole and is burnt. The hot gases leaving the grate move along the furnace (flue) tubes upto the back end of the shell and then in the downward direction to the bottom flue. The bottom of the shell is thus first heated.

The hot gases, passing through the bottom flue, travel upto the front end of the boiler, where they divide into two streams and pass to the side flues. This makes the two sides of the boiler shell to become heated. Passing along the two side flues, the hot gases travel upto the back end of the boiler to the chimney flue. They are then discharged into the atmosphere through the chimney.

With the help of this arrangement of flow passages of hot gases, the bottom of the shell is first heated and then its sides. The heat is transferred to water through the surface of the two flue tubes (which remain in water) and bottom and sides of the shell. The arrangement of flues increases the heating surface of the boiler to a large extent. Dampers control the flow of hot gases and regulate the combustion rate as well as steam generation rate.

The boiler is fitted with necessary mountings. Pressure

gauge and water level indicator provided at the front. Safety valve, steam stop valve, low water and high steam safety valve and manhole are provided on the top of the shell.

Cochran boiler

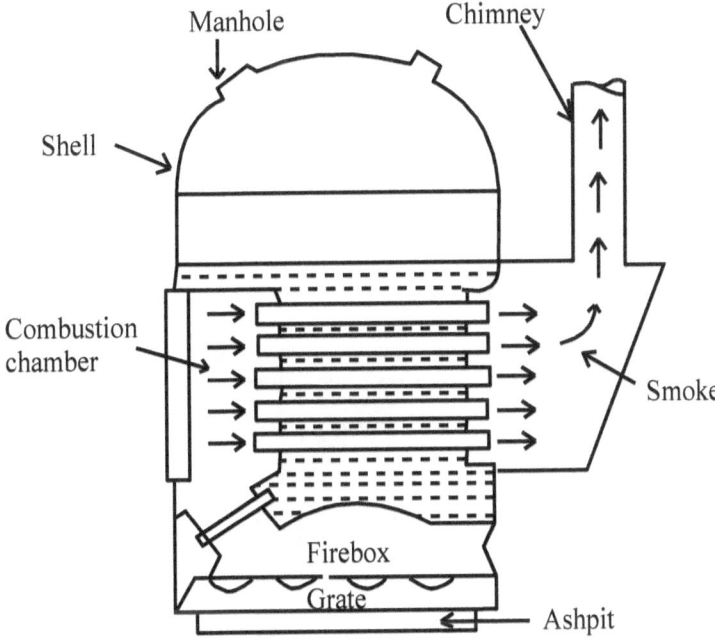

Fig.3.5: Construction of Cochran boiler

Construction: Cochran boiler consists of a cylindrical shell with a dome shaped top where the space is provided for steam. The furnace is one piece construction and is seamless. Its crown has a hemispherical shape and thus provides maximum volume of space.

Working: The fuel is burnt on the grate and ash is collected and disposed of from ash pit. The gases of combustion produced by burning of fuel enter the combustion chamber through the flue tube and strike against fire brick lining which directs them to pass through number of horizontal tubes, being surrounded by water.

29

After which the gases escape to the atmosphere through smoke box and chimney.

Advantages:

1. The minimum floor area is required.

2. Cost of construction is low.

3. It can be moved and stet up take it to different location.

4. Boiler has self-contained furnace. No brick work setting is necessary.

5. Any type of flue can be used

Water tube boilers

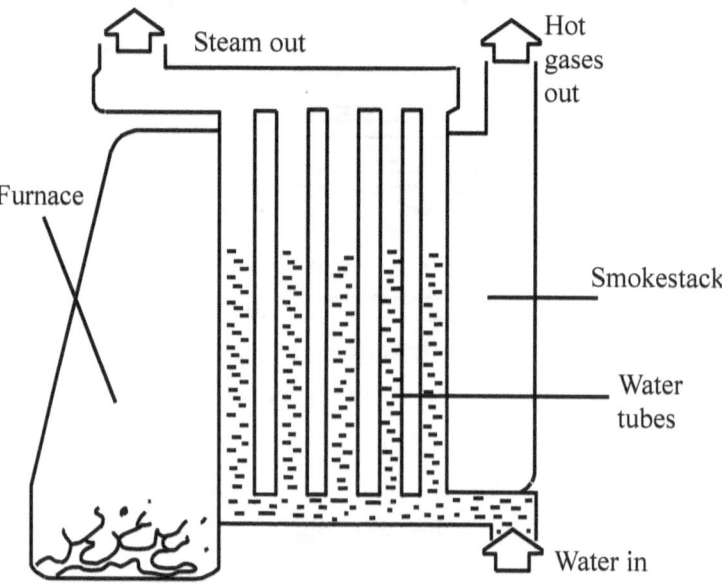

Fig.3.6: Water tube boiler

Water tube or "Water in tube" boilers are the one in which the conditions are reversed with the water passing through the tubes and hot gases passing outside the tubes. These boilers can be built to any

30

steam capacities and pressures, and have higher efficiencies than fire- tube boilers.

As the name implies, water circulates within the tubes of a water tube boiler. These tubes are often connected between two or more cylindrical drums. In some boilers the lower drum is replaced with a tube header. The higher drum, called the steam drum, is maintained approximately half full of water. The lower drum is filled with water completely and is the low point of the boiler. Sludge that may develop in the boiler gravitates to the low point and can be drawn off the bottom of this lower drum, commonly called the mud drum.

Because watertube boilers can be easily designed for greater or lesser furnace volume using the same boiler convection heating surface, watertube boilers are particularly applicable to solid fuel firing. They are also applicable for a full range of sizes and for pressures from 50 psig to 5000 psig. The present readily available minimum size of industrial watertube boilers is approximately 20,000 to 25,000 lbs/hr of steam-equivalent to 600 to 750 BoHP (boiler horsepower). Many watertube boilers operating today are in the 250 to 300 BoHP size range.

Babcock and Wilcox boiler

Construction: Babcock and Wilcox boiler with longitudinal drum. It consists of a drum connected to a series of front end and rear end header by short riser tubes. To these headers are connected a series of inclined water tubes of solid drawn mild steel. The angle of inclination of the water tubes to the horizontal is about 15° or more.

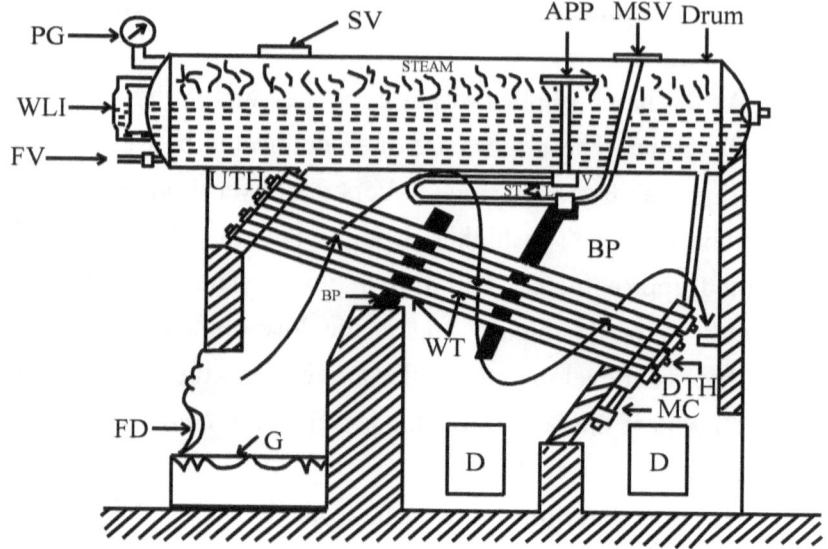

Fig. 3.7: Babcock and Wilcox boiler

DTH = Down take header; WT= Water tubes; BP= Baffle plates; D= Doors; G= Grate; FD= Fire door; MC= Mud collector; WLI= Water level indicator; PG= Pressure gauge; ST= Superheater tubes; SV= Safety valve; MSV= Main stop valve; APP= Antipriming pipe; L= Lower junction box; U= Upper junction box; FV= Feed valve.

Working: The fire door the fuel is supplied to grate where it is burnt. The hot gases are forced to move upwards between the tubes by baffle plates provided. The water from the drum flows through the inclined tubes via down take header and goes back into the shell in the form of water and steam via uptake header. The steam gets collected in the steam space of the drum. The steam then enters through the antipriming pipe and flows in the superheater tubes where it is further heated and is finally taken out through the main

32

stop valve and supplied to the Steam turbine or Steam engine when needed.

The pressure of steam in case of cross drum boiler may be as high as 100 bar and steaming capacity upto 27000 kg/h. At the lowest point of the boiler is provided a mud collector to remove the mud particles through a blow-down-cock.

Advantages

1. It uses both solid as well as liquid fuel for burning.
2. The drought losses as compared to other boiler in minimum.
3. As compared to other boiler the evaporation capacity is high.
4. The circulation of water is natural.
5. The defective tubes can be replaced easily.
6. It is used in power stations for generation large quantity of steam.

Merits and demerits of FTB over WTB

Merits

i. Have greater reliability and low fire cost because of simple and rigid construction.

ii. Due to large cylindrical drums of the fire tube boilers, there is ample water surface from which the steam can be quickly raised.

iii. Excellent for engines operating with rapid changes in load like locomotive boilers.

iv. Very compact and space occupied per kg of steam generation by fire tube boiler is considerably less than the water tube boiler.

v. Failure in feed water supply for some time does not cause damage to the boiler.

Demerits

i. Makes slow in reaching the operating pressure.

ii. Large diameter of the shell (2.4m) and limit of maximum thickness (3cm) and stress consideration limit the generation of pressure to 20 Kg/cm^2.

iii. The maximum generating capacity of these boilers is about 9000 Kg/h.

iv. The explosion of the fire tube boilers become very serious because of its large water capacity.

v. The transportation of fire tube boilers is very inconvenient because of large size of the shell.

Boiler mountings

Different fittings and devices necessary for the operation and safety of a boiler are known as boiler mountings.

Safety valve

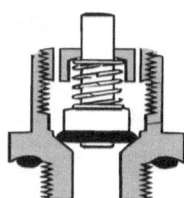

When the steam pressure in the boiler exceeds the working pressure, it lifts the valve with its weight and allows the steam to escape from the boiler to the atmosphere till the pressure reduces to the working pressure.

Fusible plug

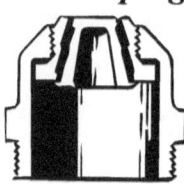

To put off the fire in the furnace of the boiler when water level in the boiler falls below an unsafe level and thus avoid the explosion which may take place due to overheating of the tubes and shell.

Water level indicator

It indicates the level of water.

Pressure gauge

To measure the pressure of the steam inside the boiler. The commonly used pressure gauge is Bourdon type pressure gauge.

Feed check valve

To allow the supply of water to the boiler at pressure continuously and to prevent the back flow of water from the boiler when the pump pressure is less than the boiler pressure or when the pump fails.

Blow- off cock

It is used for dual functions:

i. To empty the boiler when necessary for cleaning, repairing and inspecting.

ii. To discharge the mud and sediments carried with the feed water and accumulated at the bottom of the boiler.

Steam stop valve

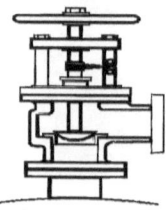

It is the largest valve on the steam boiler and usually fitted to the highest part of the shell. The function of the stop valve is to regulate the flow of steam the boiler to the engines as per requirement and shut-off the steam flow when not required.

Anti- priming pipe

Since the steam is in contact with water, it is wet and carries certain amount of moisture along with it. To reduce the moisture carry over, an anti-priming pipe with number of holes is fixed at the exit of steam from the boiler.

Boiler accessories

Boiler accessories are the auxiliary parts required for the smooth operation of the boiler and to increase the overall efficiency of the plant. Water feeding equipment, air- preheater, economizer and superheater are some of the essential accessories of the boiler.

Various boiler accessories are:

1) Air Preheater
2) Economizer
3) Superheater
4) Feed Pump
5) Injector

Economizer

The economizer usefully extracts the waste heat of the chimney gases to preheat the water before it is fed into the boiler.

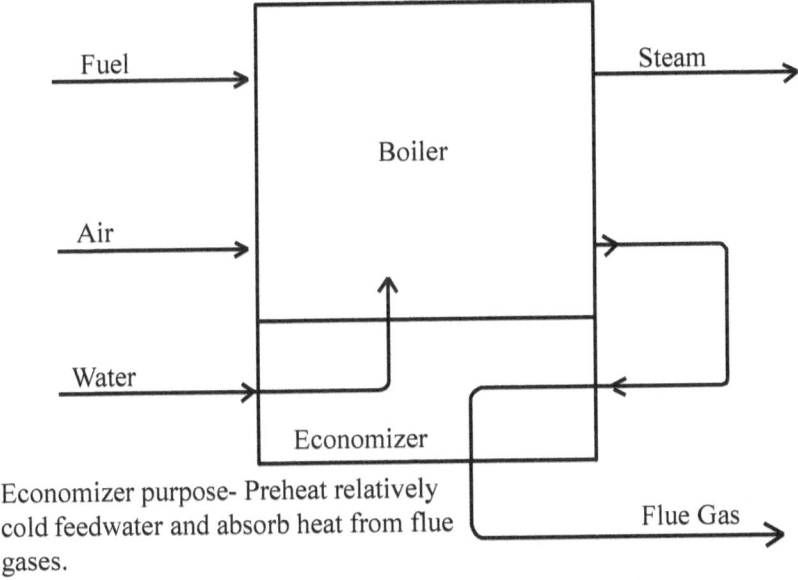

Economizer purpose- Preheat relatively cold feedwater and absorb heat from flue gases.

Fig. 3.8: A simple boiler plus economizer

Preheating of the boiler feed water has the following advantages:

i. Fuel saving as waste heat from the flue gas is used for heating the feed water.

ii. Dissolved gases as air and carbon dioxide are removed by preheating the feed water reducing corrosion and pitting.

iii. There will be less temperature strain in the boiler plates as the feed water enters the boiler at higher temperature.

iv. Circulation of water is very well maintained as quick evaporation is possible because of hot feed water.

v. This unit improves the overall efficiency of the boiler by reducing the fuel consumption.

Super heater

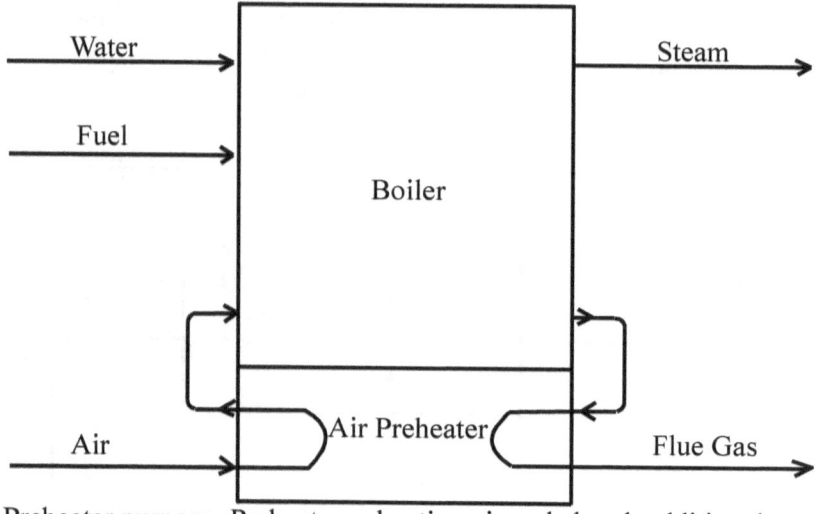

Air Preheater purpose- Preheat combustion air and absorb additional heat from flue gases.

Fig. 3.9: A simple boiler plus combustion air preheater

They are used in boilers to increase the temperature of the steam above the saturation temperature. This is done by passing the steam through a small set of tubes and hot gas over then.

Fuels used in generating steam

A fuel may be defined as a substance which produces heat when burned in the presence of air (O_2). The amount of heat generated per Kg of fuel during complete combustion is known as calorific value of fuel.

Carbon and hydrogen are the main constituent of the fuel which produce heat. The fuels, which contain mostly carbon and hydrogen as main constituents are known as hydrocarbon fuel.

Classification of fuels

Natural fuels: Solid (Wood, Coal), liquid (Crude petroleum) and gases.

Prepared fuels: Solid (Charcoal, liquid (alcohol, diesel) and gaseous.

Advantages and disadvantages of liquid fuels
Advantages of liquid fuels

i. Excess air required for complete combustion is considerably reduced as uniform mixing of fuel and air is possible due to fuel atomization.

ii. Uniform mixing of fuel and air gives high combustion efficiency and high temperature in the combustion chamber.

iii. The storage and handling of oil fuel are easy compared with coal. Therefore it reduces the handling costs.

iv. The burning rate can be easily controlled.

v. The storage required is considerably less compared to coal (less than 40%).

vi. There is no problem of hot ash disposal.

vii. The operational labour requirement is less.

viii. Oil based combustion systems are clean compared to coal fired systems.

Limitations of liquid fuels

i. The cost of liquid fuels is much higher than coal (4-5 times).

ii. Danger of explosion during transfer and storage is high.

iii. Special arrangements are required for their storage.

iv. Its use is totally dependent on its availability in the country.

Advantages and disadvantages of gaseous fuels

Advantages

i. The length and speed of the flame and the temperature in the chamber are easily controlled.

ii. They are free of solid and liquid impurities and therefore burn completely.

iii. They do not produce ash or smoke.

iv. Excess air required for complete combustion leading to better efficiency.

v. Transport of gaseous fuel through pipes is easier than the transport of coal and oil.

Disadvantages

i. Combustion system must be located near the source of gas supply or production.

ii. Highly inflammable.

Calorific value of fuels

The calorific value of fuel is the standard heat of reaction at constant pressure where the fuel burns completely with oxygen. It can be defined as the amount of heat liberated in KJ or Kcal by the complete combustion of 1 Kg of fuel.

There are 2 types of calorific values: High calorific value (HCV) and Low calorific value (LCV).

HCV

It is the total heat liberated in KJ or Kcal by the complete combustion of 1 Kg of fuel.

40

For fuels containing hydrogen, water is a product of combustion and is always formed as steam. If we allow the steam to condensate to water then we will obtain a higher heat output/ unit mass of fuel then we did not allow it to condensate. This measure of the calorific value is called HCV or Gross.

LCV

If the steam does not condensate (and this is more usually the case), we obtain the LCV or net i.e. the latent heat of steam is unavailable. It is the difference of HCV and the heat absorbed by water vapours.

$HCV = LCV + m.h_f$

Where, m = mass of H_2O produced/ kg of fuel.

h_f = enthalpy of evaporation of water at the standard temperature of 25 °C (2442.5 KJ/Kg).

Table 3.1: Some typical values of HCV and LCV (KJ/Kg)

Fuels	HCV	LCV
Wood	15.826	14.319
Petrol	46.892	43.710
Diesel	45.971	43.166
Anthracite coal	34.583	33.913

Boiler efficiency

The boiler efficiency is the ratio of the heat output to the caloric value of the fuel. Boiler efficiency is determined by various factors including the type of fuel used, the method of firing, and the control settings.

Boiler Efficiency (ɳ):

$$\eta = \frac{\text{Heat Added to Incoming Feedwater}}{\text{Heat Input (Fuel)} + \text{Heat Input (Combustion Air)}} \quad \dots\dots\dots (3.2)$$

Chapter 4: Psychrometry

Introduction

Psychrometrics is the science of studying the thermodynamic properties of moist air. The properties of air- water vapour mixture (moist air) are important in many industrial processes such as those in the paper and textile industries, air- conditioning operations for comfort in buildings, design of cooling towers and cold storage, etc.

Psychrometry is used in many industries in addition to building services engineering, e.g. aeronautical engineering, agriculture, industrial drying of crops and pharmaceuticals, food technology, meteorology and others.

Definitions of Air

Three basic definitions are used to describe air under various conditions:

Atmospheric air

It contains nitrogen, oxygen, carbon dioxide, water vapour, other gases, and miscellaneous contaminants such as dust, pollen, and smoke. This is the air we breathe and use for ventilation.

Dry air

It exists when all of the contaminants and water vapour have been removed from atmospheric air. By volume, dry air contains about 78 percent nitrogen, 21 percent oxygen, and 1 percent other gases. Dry air is used as the reference in psychrometrics.

Moist air

It is a mixture of dry air and water vapour.

Psychrometric chart

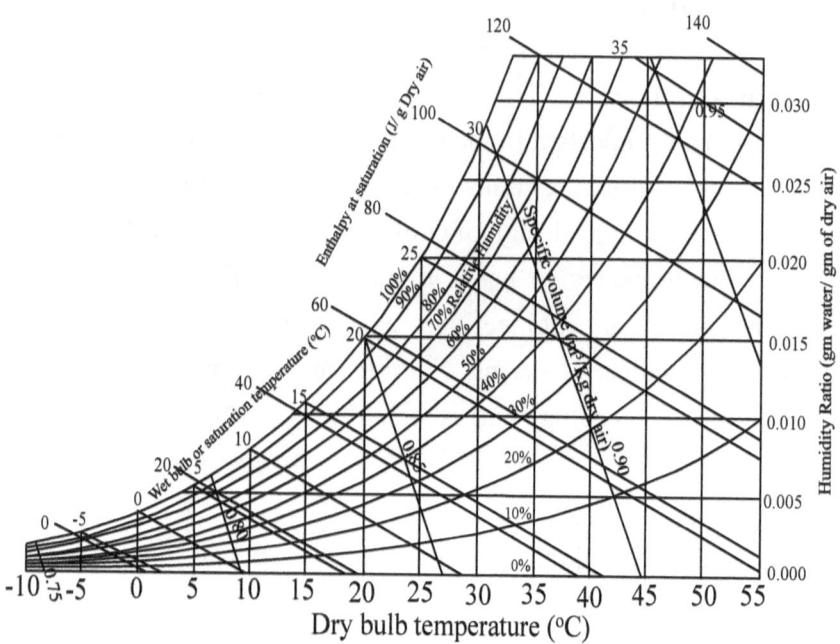

Fig. 4.1: Psychrometric chart

A psychrometric chart is a graph of the thermodynamic parameters of moist air at constant pressures, often equated to an elevation relative to sea level. A psychrometric chart is a graphical plot with dry bulb temepratures as the abscissa (Horizontal axis) and specific humidity and vapour pressure as ordinates.

The psychrometric chart is very useful during system design as a modelling design tool to enable options to be explored and to investigate system performance under different conditions. Because the chart shows visually the changes that air goes through as it is heated, cooled, mixed, humidified or dehumidified, it is easy to explore changes and control requirements. In its simplest form the psychrometric chart can show what processes are needed, what

44

equipment is required to carry them out, and how to get from a design external condition to a desired room condition.

The Psychrometric Chart provides a graphic relationship of the state or condition of the air at any particular time. It displays the properties of air: dry bulb temperature (vertical lines), wet bulb temperature (lines sloping gently downward to the right), dew point temperature (horizontal lines), and relative humidity (the curves on the chart).

Given any two of these properties, the other two can be determined using the chart. The chart's usefulness lies beyond the mere representation of these elementary properties –it also describes the air's moisture content (far right scale), energy content (outer diagonal scale on upper left), specific volume (lines sloping sharply downward to the right), and more.

Uses

i. In the design of cooling towers, condensers and air- conditioners.

ii. In the mixing of air streams, heating of air, cooling of the air, humidifying and dehumidifying of air and combination of all processes for various and industrial purposes.

iii. A psychrometric chart also helps in calculating and analysing the work and energy transfer of various air-conditioning processes.

Parameters of psychrometric chart

Dry- bulb temperature (DBT)

It is the temperature recorded by a thermometer which is not affected by moisture or radiation. Thus, such a thermometer must have no moisture and must have dry surface. It is simply the equilibrium temperature of the mixture indicated by an ordinary thermometer. In other words, the actual temperature of gas or mixture of gases indicated by an error- free temperature measuring device.

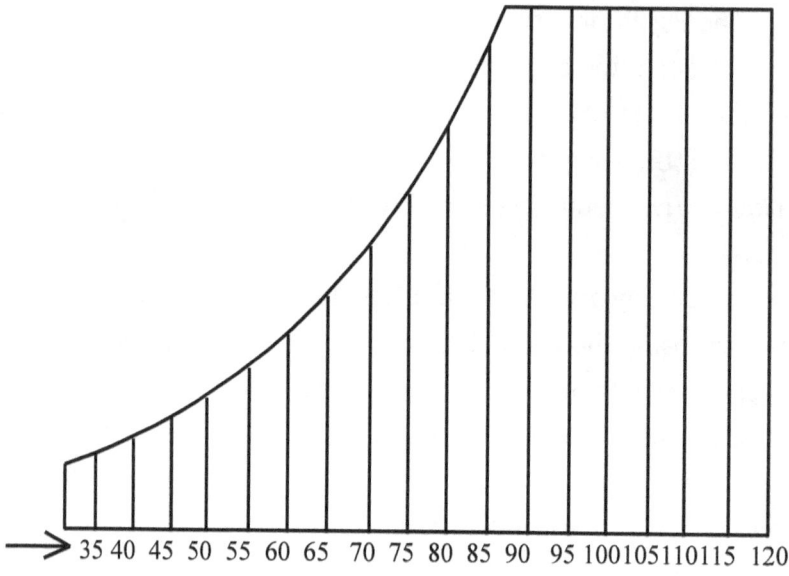

Fig. 4.2: Lines of DBT

Dry-bulb temperature is located on the X-axis, of the psychrometric chart and lines of constant temperature are represented by vertical chart lines.

Wet- bulb temperature (WBT)

As the name implies, it is the air temperature measured by a thermometer with a bulb covered with a muslin cloth sleeve and kept moist with distilled or clean water into which the air stream across the wet bulb flows with a low velocity (around 5m/s).

Wet bulb temperatures are always lower than dry bulb temperatures and the only time that they will be the same is at saturation (i.e. 100% relative humidity).

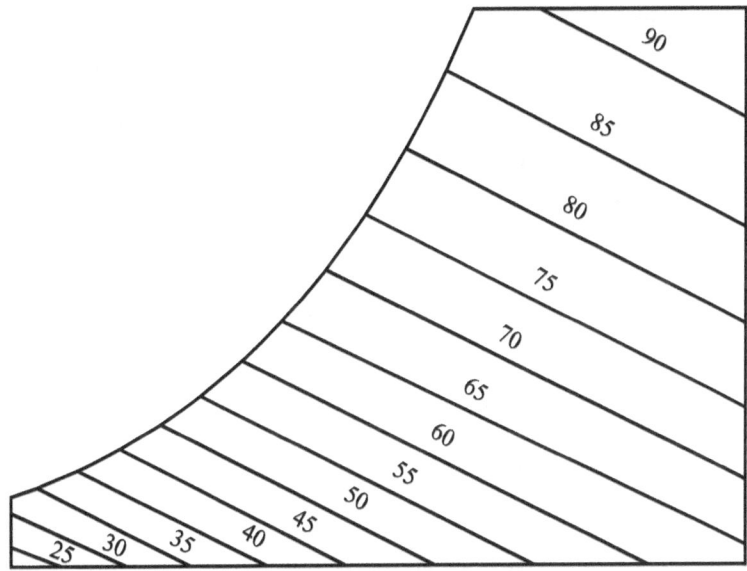

Fig. 4.3: Lines of WBT

Wet bulb temperature on psychrometric chart is represented by lines that slant diagonally from the upper right of the chart (along the line of saturation) down to the lower left of the chart. These follows lines of constant enthalpy but values are read off at the upper, curved, saturation temperature boundary.

Specific Enthalpy (H)

It is a measure of energy, but although in psychrometry, it is only the heat energy which is of interest, this is still referred to as enthalpy.

The specific enthalpy (H) is a measure of the heat energy of 1 Kg of dry air plus its associated water vapour relative to 0°C and zero moisture content.

Enthalpy is the measure of heat energy in the air due to sensible heat or latent heat.

47

Sensible heat is the heat (energy) in the air due to the temperature of the air and the latent heat is the heat (energy) in the air due to the moisture of the air. The sum of the latent energy and the sensible energy is called the air enthalpy.

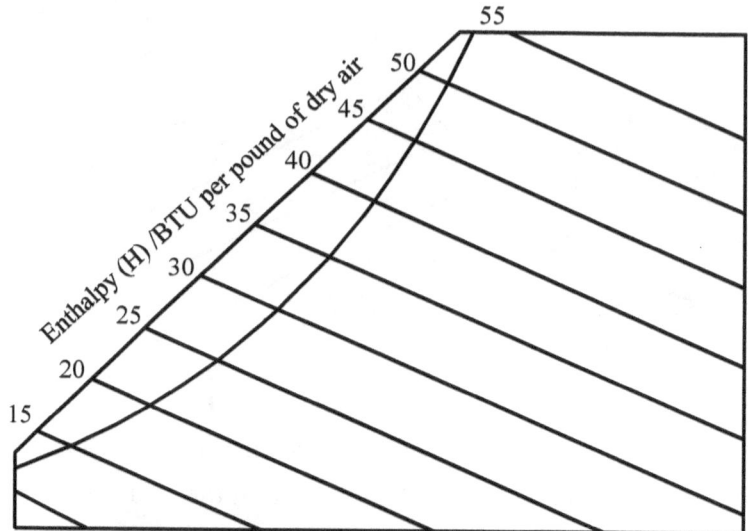

Fig. 4.4: Lines of Specific enthalpy

The Enthalpy scale is located above the saturation, upper boundary of the chart. Lines of constant enthalpy run diagonally downward from left to right across the chart; follow almost exactly the line of constant wet bulb temperature.

Relative Humidity (RH)

Relative humidity (RH) is a measure of the amount of water air can hold at a certain temperature.

It is the ratio of the vapour pressure of an air sample to the saturated vapour pressure at the same temperature and is usually expressed as percentage.

RH (%) is expressed by the equation:

$$RH = \frac{\text{Partial pressure of water vapour in the sample}}{\text{Partial pressure of water vapour in the saturated air}} \times 100\%$$

$$= \frac{P_v}{P_s} \times 100\% \dots\dots\dots (4.1)$$

As a rule of thumb, the maximum amount of water that the air can hold doubles for every 20°F increase in temperature.

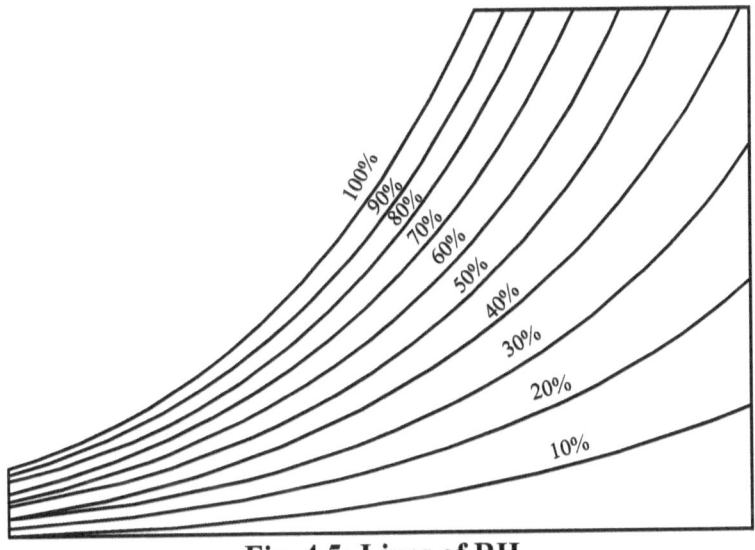

Fig. 4.5: Lines of RH

Lines of constant relative humidity are represented by the curved lines running from the bottom left and sweeping up through to the top right of the chart. The line for 100 percent relative humidity, or saturation, is the upper, left boundary of the chart.

Absolute Humidity or Humidity Ratio (W)

It is simply the ratio of the mass of water vapour to the mass of the dry air in a certain volume of mixture.

Humidity ratio is given by the equation:

49

$$W = \frac{\text{Mass of water vapour in Kg}}{\text{Mass of dry air in Kg}} = \frac{m_v}{m_a} \quad \cdots\cdots \quad (4.2)$$

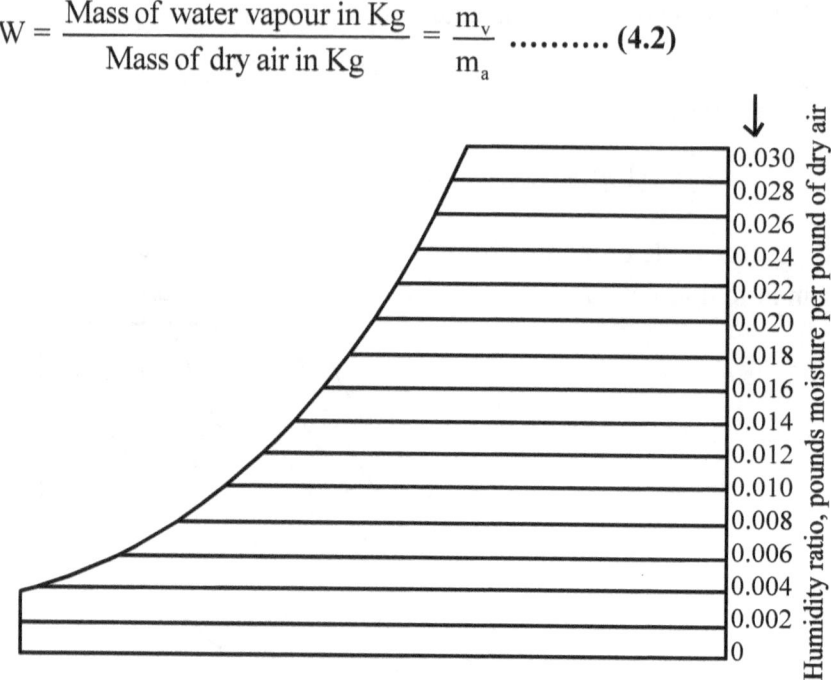

Fig. 4.6: Lines of Humidity ratio

Humidity ratio is represented on the chart by lines that run horizontally and the values are on the right hand side (Y-axis) of the chart increasing from bottom to top.

Dew Point temperature (DPT)

Dew point temperature indicates the temperature at which water will begin to condense out of moist air. When air is cooled, the relative humidity increases until saturation is reached and condensation occurs. Condensation occurs on surfaces which are at or below the dew point temperature.

Dew point is represented along the 100% relative humidity line on the psychrometric chart.

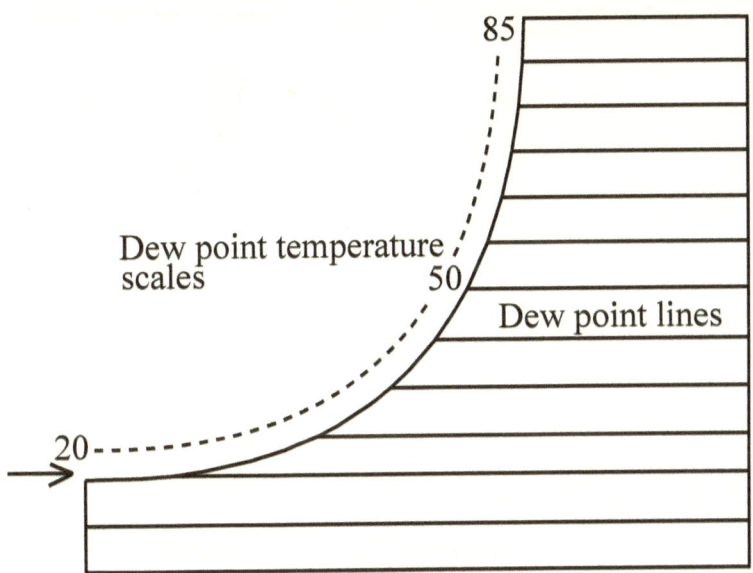

Fig. 4.7: Lines of DPT

Dew point temperature is determined by moving from a state point horizontally to the left along lines of constant humidity ratio until the upper, curved, saturation temperature boundary is reached. At dew point, dry bulb temperature and wet bulb temperature are exactly the same.

Specific Air Volume

Specific Volume is the volume that a certain weight of air occupies at a specific set of conditions. The specific volume of air is basically the reciprocal of air density. As the temperature of the air increases, its density will decrease as its molecules vibrate more and take up more space (as per Boyle's law). Thus the specific volume will increase with increasing temperature.

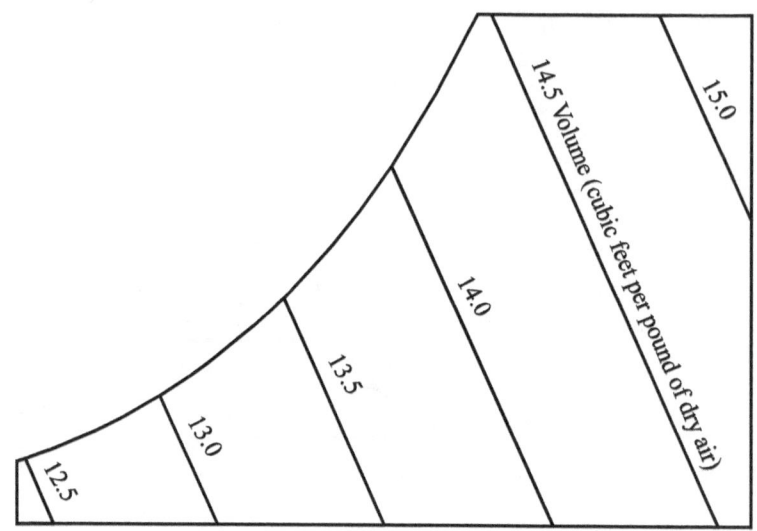

Fig. 4.8: Lines of specific air volume

Specific volume is represented on Psychrometric Chart by lines that slant from the lower right hand corner towards the upper left hand corner at a steeper angle than the lines of wet bulb temperature and enthalpy.

Example 4.1: Moist properties of air

1. A sling psychrometer gives a dry-bulb temperature of 78°F and a wet-bulb temperature of 65°F. Determine other moist air properties from this information.

Sol^n:

i. Use psychrometric chart measured at sea level.

ii. You can convert the units of temperature scale to degree centigrade if it's not compatible to the one in the chart available.

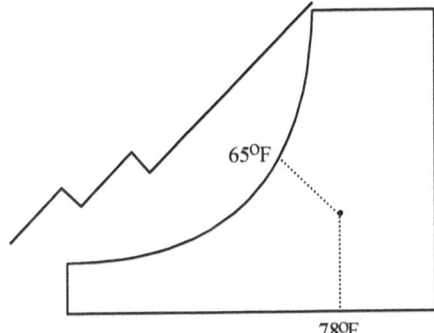

1. Find the intersection of the two known properties, dry-bulb and wet-bulb temperatures, on the psychrometric chart.

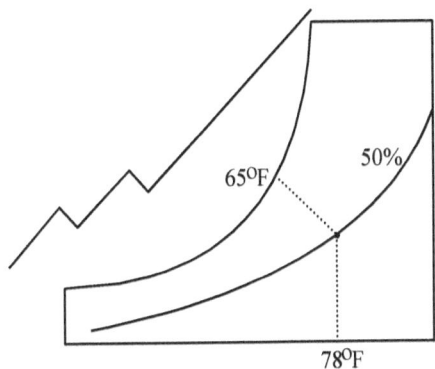

2. RH = 50%

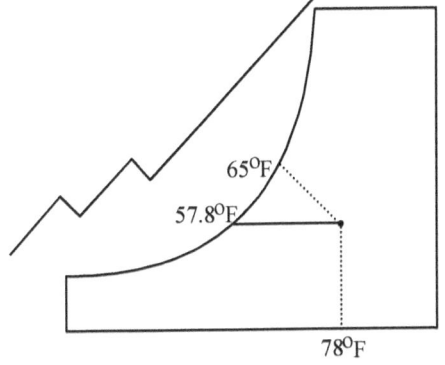

3. Dew Point = 57.8 degrees

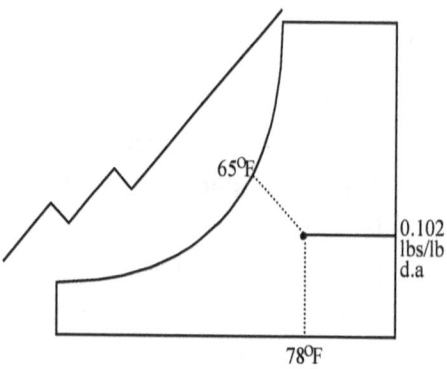

4. Absolute humidity = .0102 lbs water / lb dry air

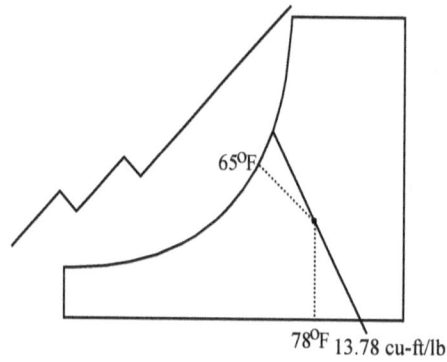

5. Specific volume = 13.78 cu-ft/lb dry air

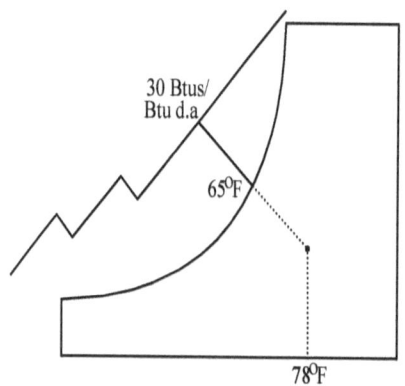

6. Enthalpy = 30 Btu's/lbs dry air

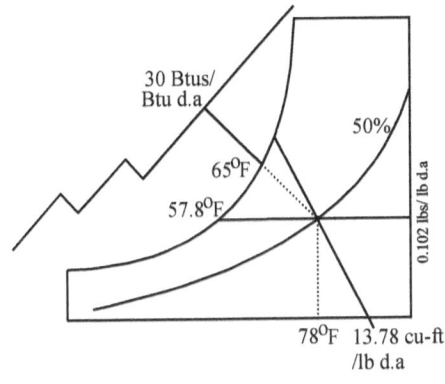

Hence,

Relative humidity = 50%

Dew Point = 57.8 degrees

Absolute humidity = .0102 lbs water / lb dry air

Specific volume = 13.78 cu-ft/lb dry air

Enthalpy = 30 Btu's/lbs dry air

Exercises on psychrometric chart

1. The dry bulb reading is 78°F and the wet bulb is 58°F. Using the chart determine the other properties of air.

Answer:

Relative humidity = 28%

Dew Point = 42 degrees

Absolute humidity = .0056 lbs water / lb dry air

Specific volume = 13.65 cu-ft/lb dry air

Enthalpy = 25 Btu's/lbs dry air

2. Find the moisture content of air at 25°C dry-bulb temperature and 25°C wet-bulb temperature. [**Ans:** 0.020 kg/kg dry air]

3. Find the specific volume and wet-bulb temperature of air at 20°C dry-bulb temperature and 50% saturation. [**Ans:** 0.84 m³/kg, 14°C]

4. Find the specific volume, percentage saturation and moisture content of air at 15°C dry-bulb temperature and 10°C wet-bulb temperature. [**Ans:** 0.823 m³/kg, 52% & 0.0054 kg/kg d.a]

5. Find the specific volume, wet-bulb temperature, moisture content and specific enthalpy of air at 35°C dry-bulb temperature and 30% saturation. [**Ans:** 0.883 m³/kg, 22°C, 0.011kg/kg d.a. & 65 kJ/kg]

Psychrometric processes

A thermodynamic system is said to undergo a process whenever it changes from one state-point to another. In psychrometric analysis this change in state can be caused by adding or removing heat, causing work to be performed on or by the fluid, adding or removing mass (usually of water vapour), or mixing two fluids which are at dissimilar states.

Sensible Heating or Cooling

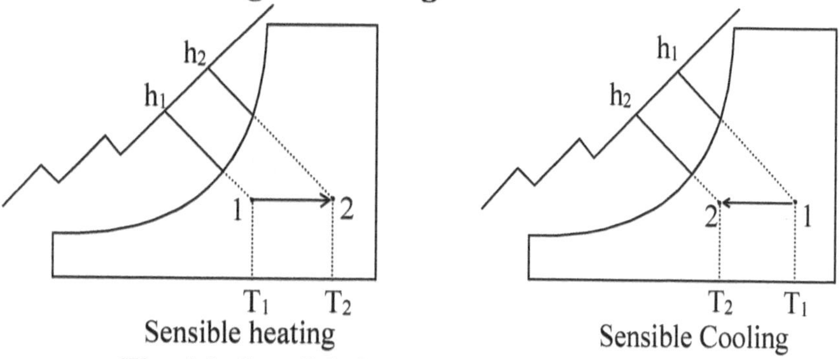

Sensible heating   Sensible Cooling

Fig. 4.9: Sensible heating and cooling process

It is a psychrometric process that involves the increase or decrease in the temperature of air without changing its humidity ratio. e.g: passing moist air over a room space heater and of kiln air over the heating coils.

Observable features of a sensible heating process

i. The dry bulb temperature increases

ii. The relative humidity decreases

iii. The enthalpy increases

56

iv. The wet bulb temperature increases

v. The specific volume increases

vi. The humidity ratio, vapour pressure and dew point remains constant

Observable features of a sensible cooling process

i. The dry bulb temperature decreases

ii. The relative humidity increases

iii. The enthalpy decreases

iv. The wet bulb temperature decreases

v. The specific volume decreases

vi. The humidity ratio, vapour pressure and dew point remains constant

Governing Equation for Sensible Heating and Cooling:

1) $Q_s = M_a*(h_2 - h_1)$.......... (4.3), and

2) $qs = m_a*(h_2 - h_1)$ .... (4.4)

Where:

- Q_s = sensible heat added, Btu
- M_a = Mass of dry air, lb [= (volume of air) /(specific volume of moist air in ft^3/lb d.a)]
- h = Enthalpy of air, Btu/lb of dry air
- q_s = Rate of sensible heat transfer, Btu/min
- m_a = Mass airflow, lb/min [= (volume of moist airflow per min)/(specific volume of moist air in ft^3/lb d.a)]

Example 4.2: Calculation of sensible heat

1. Calculate the amount of sensible heat that must be added to 100lb of air at 85°F dry bulb and 75°F wet bulb to raise the temperature of air to 100°F dry bulb.

Sol^n:

1. Locate the 85°F dry bulb and 75°F wet bulb – this corresponds to 62% Relative humidity.

2. Move from point established in a) above to right to 100°F on a horizontal dry bulb temperature scale – this corresponds to 39.6% Relative humidity.

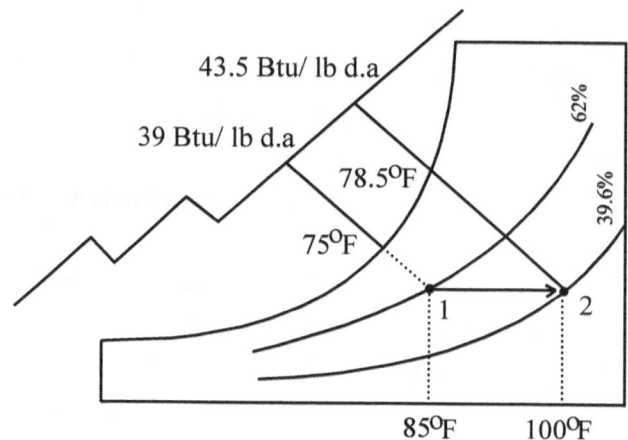

The condition of the air is changing from state 1 to state 2:

Read h_2 and h_1 from psychrometric chart as:

h_2 = 39 Btu's/lb dry air

h_1 = 43.5 Btu's/lb dry air

Estimate heat addition:

$Q = m [h_2 - h_1] = 10 \times [43.5 - 39] = 450$ BTU's.

Hence, the amount of sensible heat required is 450 BTU.

Exercises on sensible heating and cooling

1. Determine the amount of sensible heat needed to increase the temperature of air from 50°F and 50% RH to 90°F. [**Ans:** 10 Btu/lb]

2. Moist air, saturated at 2°C, enters a heating coil at a rate of 10 m³/s. Air leaves the coil at 40°C. Find the required rate of heat addition. [**Ans: 492 KW**]

Heating and Humidifying

It is a psychrometric process that involves the simultaneous increase in both the dry bulb temperature and humidity ratio of the air.

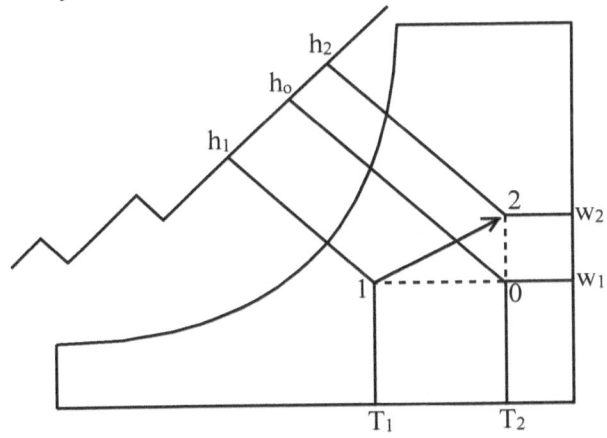

Fig. 4.10: Heating and humidification process

  The heating and humidification of the air is best considered by looking at the two processes sequentially.

  The first, from state 1 to state 0, is the sensible heating that occurs when the air passes through the heat exchanger. The second, from state 0 to state 2, is the humidification process.

Observable features of a humidification process

i. Humidification adds moisture to the air which increase the absolute humidity

ii. Water is added in vapour form

iii. Water is converted from liquid to gas

iv. There is an increase in the energy level

Governing Equation for Heating and Humidification:

If the condition of the air is changing from state 1 to state 2, consider that the intersection of a horizontal line through state 1 and a vertical line through state 2 occurs at state 0. Then the change in heat can be expressed as:

1) $Q_s = M_a*(h_0 - h_1)$ and $q_s = m_a*(h_0 - h_1)$ .... **(4.5)**

2) $Q_L = M_a*(h_2 - h_0)$ and $q_L = m_a*(h_2 - h_0)$ .... **(4.6)**

Where:

- Q_s = sensible heat added, Btu
- Q_L = latent heat added, Btu
- M_a = Mass of dry air, lb [= (volume of air) /(specific volume of moist air in ft^3/lb d.a)]
- h = Enthalpy of air, Btu/lb of dry air
- q_s = Rate of sensible heat transfer, Btu/min
- m_a = Mass airflow, lb/min [= (volume of moist airflow per min)/(specific volume of moist air in ft^3/lb d.a)]

The rate of moisture addition to the air, M_{water}, is determined by a water vapour mass balance:

3) $M_{water} = M_a (w_2 - w_0)$ or $m_w = m_a (w_2 - w_0)$ .... **(4.7)**

Where:

- w_2 = humidity ratio of the moist air upstream of the humidifier
- w_0 = humidity ratio of the moist air downstream of the humidifier

Example 4.3: Drying of lumber

1. Ninety cubic-ft of lumber is dried at 60°C (140°F) dry bulb temperature and 52°C (125.6°F) wet bulb temperature. The drying rate of the lumber is 5.68 lb of water per hour. If outside air is at 27°C (80.6°F) dry bulb temperature and 80% relative humidity, how

much outside air is needed per minute to carry away the evaporated moisture?

Sol^n:

i. Locate the 80.6°F dry bulb and 80% Relative humidity as Node 1

ii. Locate the 140°F dry bulb and 125.6 °F wet bulb temperature as Node 2.

iii. Move from point established in a) above to right to 100°F on a horizontal dry bulb temperature scale – this corresponds to 39.6% Relative humidity.

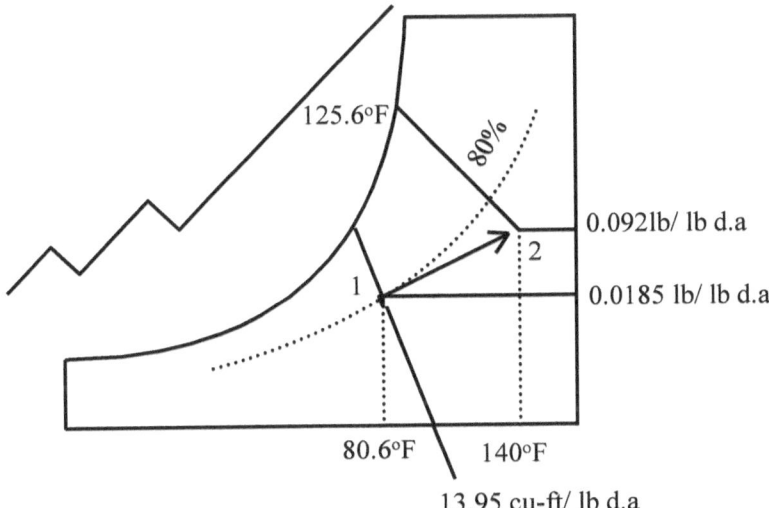

The condition of the air is changing from state 1 to state 2:

Read absolute humidity w_2 and w_1:

• $w_2 = 0.092$ lb/lb dry air

• $w_1 = 0.0185$ lb/lb dry air

Estimate moisture gain : $\Delta HR = [w_2 - w_1]$ lb/lb dry air

• $\Delta HR = (0.092 - 0.0185)$ lb/lb dry air

• $\Delta HR = 0.0735$ lb/lb dry air

Estimate drying air: w = drying rate/ΔHR

• w = (5.68 lb/hour)/ (0.0735 lb/lb dry air) = 77.28 lb dry air/hour

Estimate volumetric air flow rate for drying: V = w x specific volume

• V = (77.28 lb dry air/hour) x (13.95 ft³/lb dry air) = **1078 ft³/hour**

Exercises on heating and humidifying

1. How much moisture is added to 20 lb of air going from 50°F, 50% RH to 80°F, 60% RH?

Answer:

HR (50 F, 50% RH) = 0.0038 lb/lb d.a

HR (80 F, 60% RH) = 0.0132 lb/lb d.a

Water added = 20 lb * (0.0132 - 0.0038) lb/lb = 0.188 lb-m

Cooling and Dehumidifying

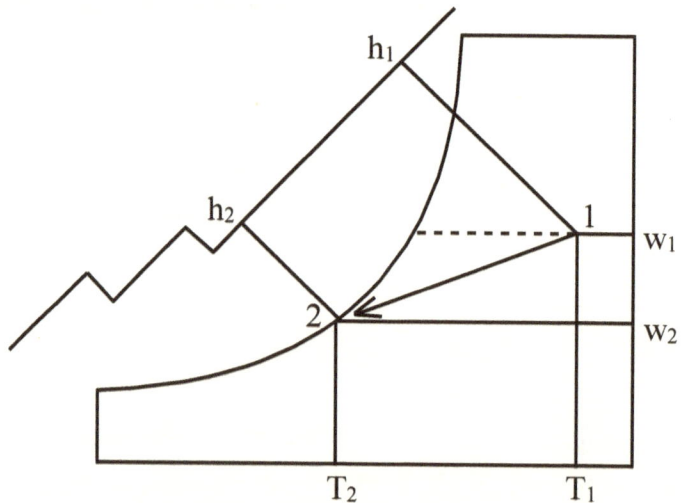

Fig. 4.11: Cooling and dehumidifying process

It is a psychrometric process that involves the removal of water from the air as the air temperature falls below the dew point temperature.

Observable characteristics of a cooling and dehumidifying

i. Dry bulb temperature decreases

ii. Humidity ratio decreases

iii. Vapour pressure decreases

iv. Dew point temperature decreases

v. Wet bulb temperature decreases

vi. Enthalpy decreases (there is a decrease in the energy level and with the loss of energy, condensation occurs)

vii. Relative humidity increases

Governing Equations:

The combined process of cooling and dehumidification can be modeled using mass balances on the air and water, and an energy balance on the entire process. The steady state mass and energy balances are shown below.

Mass balance (air): $m_{a1}{}^* = m_{a2}{}^*$

Mass balance (water): $m_{w1}{}^* = m_{w2}{}^* + m_w{}^*$

Or, $m_w{}^* = m_{w1}{}^* - m_{w2}{}^* = m_{a1}{}^* w_1 - m_{a2}{}^* w_2 = m_a{}^*[w_1-w_2]$

Energy balance: $m_{a1}{}^* h_1 - Q_c - m_a{}^* h_2 - m_w{}^* h_w = 0$

$$\text{Or, } Q_c = m_a{}^*[h_1 - h_2] - m_w{}^* h_w$$

Example 4.4: Dehumidification of air

1. Consider a hot humid day 90°F and 90% RH. We want to condition the air to 70°F at about 50% RH. We do this by chilling the air far enough to condense out enough moisture to dehumidify it: the goal is to have air with absolute humidity not exceeding 0.008 lbs of moisture per pound of air (~50 to 55 grains per pound of dry air). Show the processes on the psychrometric chart.

Sol^n:

i. Plot 90° and 90% RH on the chart

ii. Read the absolute humidity at this point to 0.029 lbs moisture per lb of dry air or (195grams of moisture per pound of air)

iii. Check the final condition of 0.008 lbs moisture per lb of dry air and run the horizontal line to saturation curve. Read the temperature as 50°F

iv. Cool the air from 90°F @ 90% RH to 50°F – Now there are 0.008 lbs moisture per lb of dry air and 100% RH. 0.021 lbs of moisture per lb of dry air (142 grains of moisture) have condensed out – the air is now dehumidified.

v. The air is dehumidified, but cold (50°F) and is at 100% RH; however it only has 53 grams of moisture.

vi. Warm Back Up to 70°F (sensible heating), the RH raises to ~50%

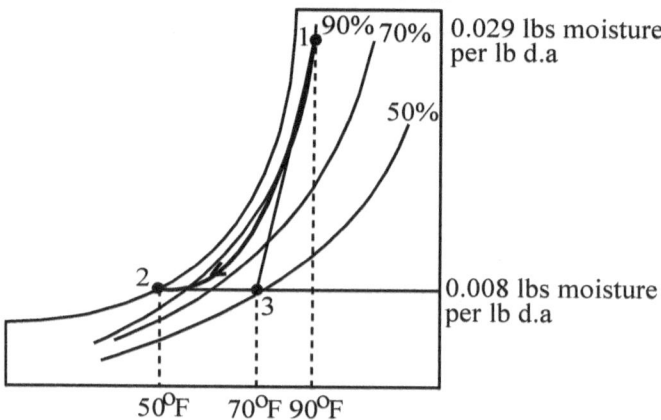

Summary

We had hot humid air

– chilled it to condense the moisture out (state point 2)

– warmed it back up to reach our target humidity (state point 3)

– Table below gives the temp, RH and specific humidity at each step.

Temperature	RH	Specific Humidity
90 °F	90%	0.029 lbs of moisture per lb d.a
50°F	100%	0.008 lbs of moisture per lb d.a
70°F	50%	0.008 lbs of moisture per lb d.a

Water activity (a$_w$)

It is very important parameter for describing various biological reactions including microbial growth. It is a measure of unbound, free water in the system available to support biological and chemical reactions.

It can simply be defined as the ratio of the partial water vapour pressure over a system (food) to that of pure water at a given temperature.

$$a_w = \frac{P}{P_o} \quad \ldots\ldots\ldots \textbf{(4.8)}$$

It is also related to relative humidity as:

$$a_w = \frac{1}{100} \times ERH \quad \ldots\ldots\ldots \textbf{(4.9)}$$

Equilibrium Relative Humidity (ERH) can simply be defined as the humidity at a given temperature where the food will neither loose moisture to the atmosphere nor pick up moisture from the atmosphere.

Most bacteria require aw in the range of about 0.90 to 1.0. Some yeasts and molds grow slowly at aw down as low as about 0.65.

Two foods with the same water content can have very different a$_w$ values depending upon the degree to which the water is free or otherwise bound to food constituents.

Microorganisms vs. a_w value

The definition of moisture conditions in which pathogenic or spoilage microorganisms cannot grow is of paramount importance to food preservation. It is well known that each microorganism has a critical a_w below which growth cannot occur. For instance, pathogenic microorganisms cannot grow at a_w <0.86; yeasts and moulds are more tolerant and usually no growth occurs at a_w <0.62. The so-called intermediate moisture foods (IMF) have a_w values in the range of 0.65-0.90 (Fig. 4.12).

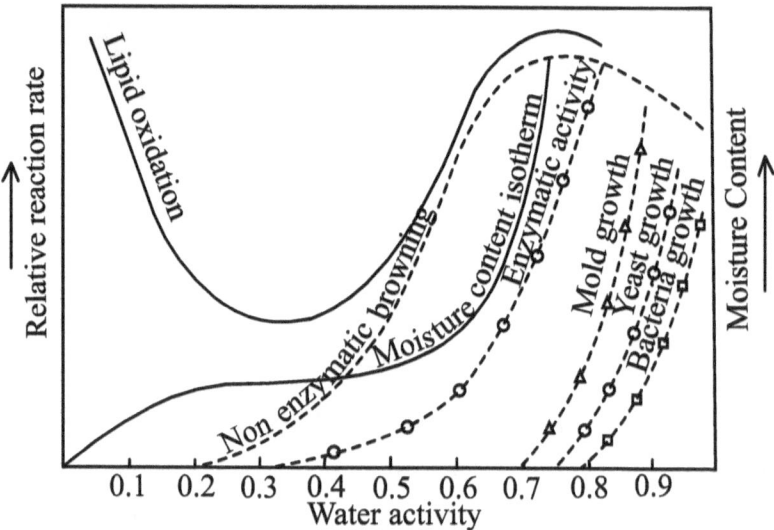

Fig. 4.12: **Water activity stability diagram**

Enzymatic and chemical changes related to a_w values

The relationship between enzymatic and chemical changes in foods as a function of water activity is illustrated in Figure 4.12. With aw at 0.3, the product is most stable with respect to lipid oxidation, non-enzymatic browning, enzyme activity, and of course, the various microbial parameters. As aw increases toward the right, the probability of the food product deteriorating increases.

According to Rahman and Labuza (1999), enzyme-catalyzed reactions can occur in foods with relatively low water contents. The authors summarized two features of these results as follows:

i. The rate of hydrolysis increases with increased water activity but is extremely slow with very low activity.

ii. For each instance of water activity there appears to be a maximum amount of hydrolysis, which also increases with water content.

The apparent cessation of the reaction at low moisture cannot be due to the irreversible inactivation of the enzyme, because upon humidification to a higher water activity, hydrolysis resumes at a rate characteristic of the newly attained water activity. Rahman and Labuza (1999) reported the investigation of a model system consisting of avicel, sucrose, and invertase and found that the reaction velocity increased with water activity. Complete conversion of the substrate was observed for water activities greater than or equal to 0.75. For water activities below 0.75, the reaction continued with 100% hydrolysis. In solid media, water activity can affect reactions in two ways: lack of reactant mobility and alternation of active conformation of the substrate and enzymatic protein. The effects of varying the enzyme-to-substrate ratios on reaction velocity and the effect of water activity on the activation energy for the reaction could not be explained by a simple diffusional model, but required postulates that were more complex:

i. The diffusional resistance is localized in a shell adjacent to the enzyme.

ii. At low water activity, the reduced hydration produces conformational changes in the enzyme, affecting its catalytic activity.

The relationship between water content and water activity is complex. An increase in aw is usually accompanied by an increase in water content, but in a non-linear fashion. This relationship between water activity and moisture content at a given temperature is called the moisture sorption isotherm. These curves are determined experimentally and constitute the fingerprint of a food system.

Importance of humidity in foods

Bacteria, yeasts and molds are those micro- organisms which attack virtually all the food constituents, when food is contaminated under natural conditions. Uncontrolled cooling and freezing deteoriate foods. If the water is removed from food, it will also transfer out bacterial cells and multiplication will stop.

Bacteria and yeasts generally require more moisture than molds, and so molds often will be found growing on semi- dry foods where bacteria and yeasts find conditions unfavourable.

In the figure below, meat was stored cold for 20 days at 75% and 95% RH. At each temperature of cold storage, it is seen that the higher humidity produced the higher population of micro-organisms.

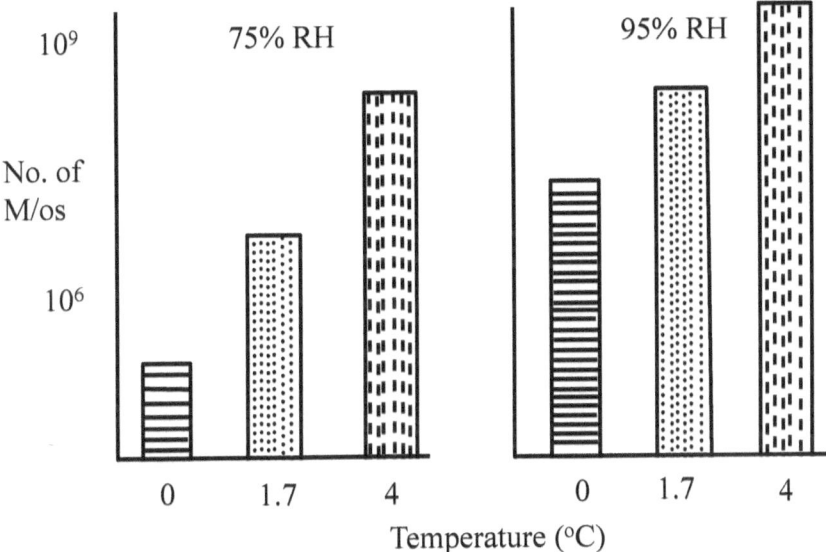

Fig. 4.13: RH and temperature vs meat quality

The RH and air motion are important factors which must be controlled in the storage of all perishable foods in their natural state. Low RH and high air velocity causes excessive dehydration in the stored product.

The losses of moisture from the fruits and vegetables causes reduction in weight and vitamins. In case of meat, it causes shrinkage and discoloration.

Hence, humidity plays a vital role in food preservation by means of controlling bacteria, yeasts and molds multiplication, growth and chemical reactions.

Chapter 5: Temperature measuring devices

Temperature
Temperature is the measure of level of heat in a substance. Thermometer is an instrument used for the measurement of temperature.

Temperature scales
i. Celsius or Centigrade scale

ii. Fahrenheit scale

iii. Absolute temperature (Kelvin scale)

iv. Absolute scale (Rankin scale)

Celsius or Centigrade scale
The melting point of ice is 0°C and the boiling point of water is 100°C. The interval between these two points is divided into 100 equal divisions and each division is called one degree Celsius (1°C).

Fahrenheit scale
According to this scale, at standard atmospheric pressure, the freezing point of ice is 32°F and the boiling point is 212 °F. The interval between these two points is divided into 180 equal divisions and each divisions is called one degree Fahrenheit (1°F).

*Absolute temperature
Absolute zero is the lowest temperature theoritically possible characterized by the complete absence of heat. Absolute zero is – 273.15 °C (-459.67°F) or zero degrees on the Kelvin scale (0°K). Absolute zero cannot be reached experimentally although it can be closely approached.

Absolute temperature (Kelvin) scale
In this scale, absolute zero is at -273.15°C which is 0K, and the degree intervals are identical to those measured on the Celsius scale. Hence, the melting point of ice on the kelvin absolute scale is (0 +273) = 273 K.

Absolute scale (Rankin) scale

It is the absolute Fahrenheit scale which places absolute zero at 0°R which is -459.67°F. Hence, the absolute temperature of melting point of ice is (32 +460) = 492 °R.

Temperature measuring devices

Thermocouple

The thermocouple is one of the simplest and most commonly used methods of measuring process temperatures.

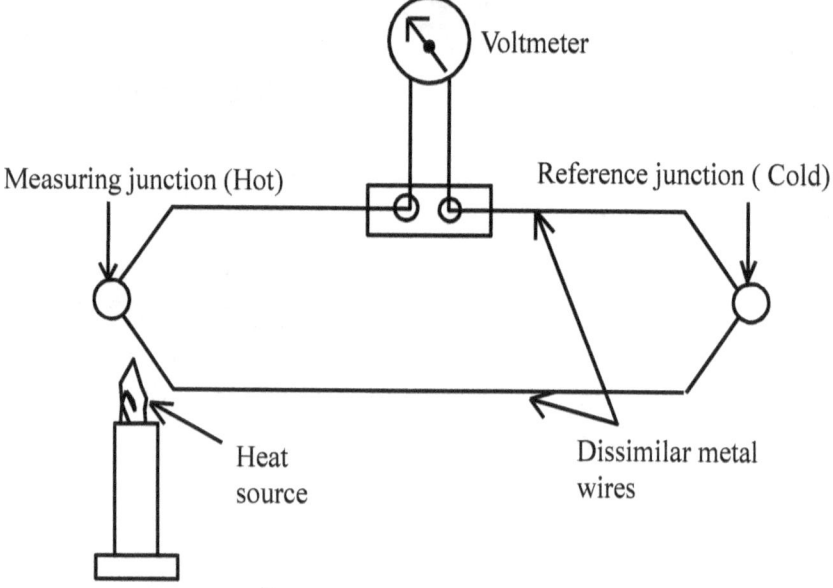

Fig. 5.1: Thermocouple

A thermocouple is an electrical device consisting of two different conductors forming electrical junctions at differing temperatures. A thermocouple produces a temperature-dependent voltage as a result of the thermoelectric effect, and this voltage can be interpreted to measure temperature. Thermocouples are a widely used type of temperature sensor.

The operation of thermocouple is based upon seebeck effect. i.e. when heat is applied to junction (Hot junction) of two dissimilar metals, then an emf is generated which can be measured at the other junction (Cold junction).

The two dissimilar metals form an electronic circuit, and a current flows as a result of the generated emf as shown in the figure. The current will continue to flow as long as $T_1 > T_2$.

Thermocouples are widely used in science and industry; applications include temperature measurement for kilns, gas turbine exhaust, diesel engines, and other industrial processes. Thermocouples are also used in homes, offices and businesses as the temperature sensors in thermostats, and also as flame sensors in safety devices for gas-powered major appliances.

Thermopile

A thermopile is an electronic device that converts thermal energy into electrical energy. It is composed of several thermocouples connected usually in series or, less commonly, in parallel. The purpose of using a thermopile rather than a single thermocouple is off course to obtain a more sensitive element.

The total output from a thermocouple connected to form a thermopile will be equal to the sum of the individual emf's and if the thermocouples are identical, the total output will be equal "n" times the output of a single thermocouple.

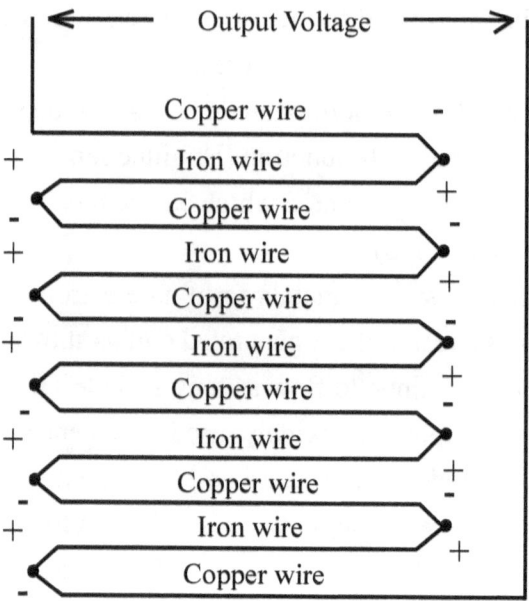

Fig. 5.2: Thermopile

A thermopile may employ 8- 16 numbers of thermocouple junction either exposed or contained in vacuum cell.

Thermistors

Thermistor is a contraction of a term "thermal resistors". Thermistors are generally composed of semiconductor materials. Most thermistors have a negative coefficient of temperature resistance i.e. their resistance decreases with increase in temperature.

Assuming as a 1st order approximation relationship between resistance and temperature is linear,

$\Delta R = K \Delta T$ where K is 1st order temperature coefficient of resistance.

74

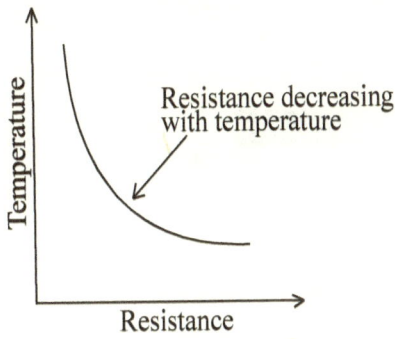

The negative temperature coefficient of resistance can be as large as several percent per degree Celsius. This allows the thermistors circuits to detect very small changes in temperature which could not be observed with a thermocouple.

As the resistance is lowered with higher temperatures, it allows more electricity to flow in predictable amounts. By measuring the electricity, we can find the temperature that correlates with the electricity flow.

Bimetallic Strip thermometer

Bimetallic strip thermometers are mechanical thermometers. They are widely used in industry for temperature control because of their robustness, temperature range and simplicity.

It consists of two strips made of dissimilar metals and bonded together with one end fixed and the other free. The principle is that as the temperature changes one strip expands more than the other, causing the pair to bend at the free end.

Most bimetallic strips use a high thermal expansion alloy, such as steel or stainless steel, coupled with a low thermal expansion alloy such as Invar.

The bimetallic strips are bonded together in a cantilever. The deflection is used to indicate temperature as shown in Fig. 5.3.

Bimetallic strips are used in thermostats for measuring and controlling temperature. The strip is connected to a switch and as the temperature changes the strip flexes and opens or closes a contact.

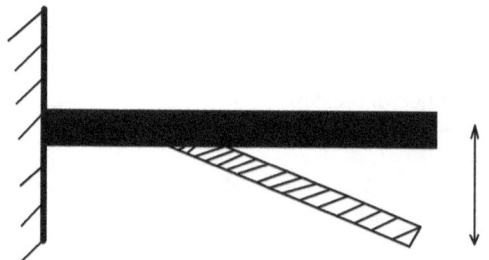

Fig. 5.3: Bimetallic strip thermometer

They are also used in ovens for measuring temperature. The strip is coupled to a dial which is calibrated for temperature indication.

Bimetallic strip advantages and disadvantages

i. Power source not required

ii. Robust, easy to use and cheap but not very accurate. Can be used to 500 °C

iii. Limited to applications where manual reading is acceptable, e.g. a household thermometer

iv. Not suitable for very low temperatures because the expansion of metals tend to be too similar, so the device becomes a rather insensitive thermometer.

Radiation pyrometers

When temperature being measured are high and physical contact with the process to be measured is impracticable or impossible, use is made of thermal radiation methods or optical pyrometers are used. These pyrometers find applications for temperatures which are above the range of thermocouples and also for rapidly moving objects.

There are 2 types of pyrometers:

i. Total radiation pyrometer

ii. Optical pyrometer.

76

Total radiation pyrometer

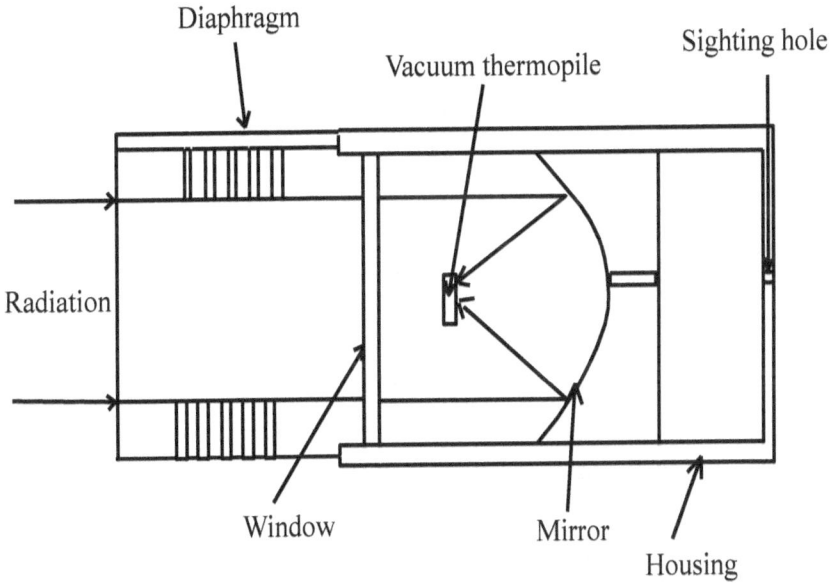

Fig. 5.4: Total radiation pyrometer

In this case, the total radiant energy from a heated body is measured. The total radiation pyrometer receives virtually all the radiant from a particular area of a hot body and focuses it on a thermocouple, thermopile, bolometer, etc. It uses lens or mirror to concentrate the radiation to surface of a thermopile.

Optical pyrometer

The most common type of optical pyrometer is the "Disappearing filament type pyrometer". The disappearing-filament pyrometer is an optical pyrometer, in which the temperature of glowing incandescent object is measured by comparing it to the light of a heated filament as shown in figure 5.5.

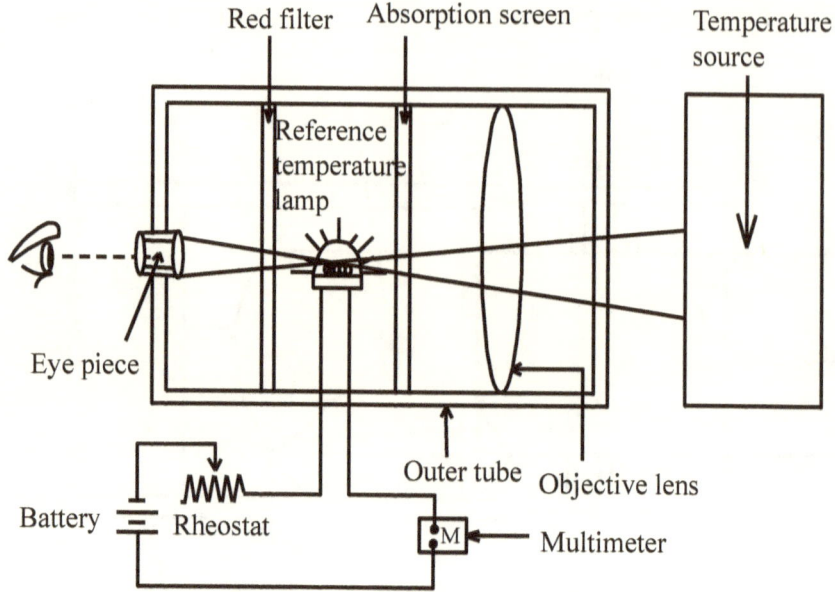

Fig. 5.5: Disappearing filament type pyrometer

The pyrometer is sighted at unknown temperature source at a distance such that the objective lens focus the source in plane of lamp filament. The current through the lamp filament is made variable so that the lamp intensity can be adjusted. The filament is viewed through an eye piece and a filter. The current through filament is adjusted until the filament and the image are of equal brightness. The brightness of the light of given colour emitted by a hot source gives an indication of temperature.

Many disappearing-filament pyrometers use a red filter. The combination of the filter and the human eye's response only allows through a narrow band of red wavelengths, so the luminosity comparison is made over only a narrow band of wavelengths. This reduces errors due to the target and filament not having identical emission spectra. For very hot objects, additional filters can be used

78

to protect the eye from excessive light. The resolution of the instrument depends somewhat on the operator, but with a skilled operator a resolution of 10°C for temperatures up to 2000°C can be achieved.

Disappearing-filament pyrometers can be used only if the object under study emits visible light similar to a hot black body; this means that its temperature must be high enough (around 600°C and up) and the object must not be fully transparent or highly reflective. For good accuracy, the object should appear dark grey or black when cold.

The optical pyrometer is widely used for accurate measurement of temperature of furnaces, molten metals and other heated materials.

Chapter 6: Pressure measuring devices

Pressure

It is the force per unit area exerted by a liquid or gas on a body or surface, with the force acting at right angles to the surface uniformly in all directions.

Pressure is measured in SI units (N/m^2) which is 1 pascal or $1 N/m^2$.

$1 Pa = 1 N/m^2 = 10^{-5}$ bar.

$1 atm = 101325 Pa = 760$ torr $= 760$ mm of Hg.

1 torr $= 1$ mm of Hg $= 133.322$ Pa.

Relation between different types of pressure

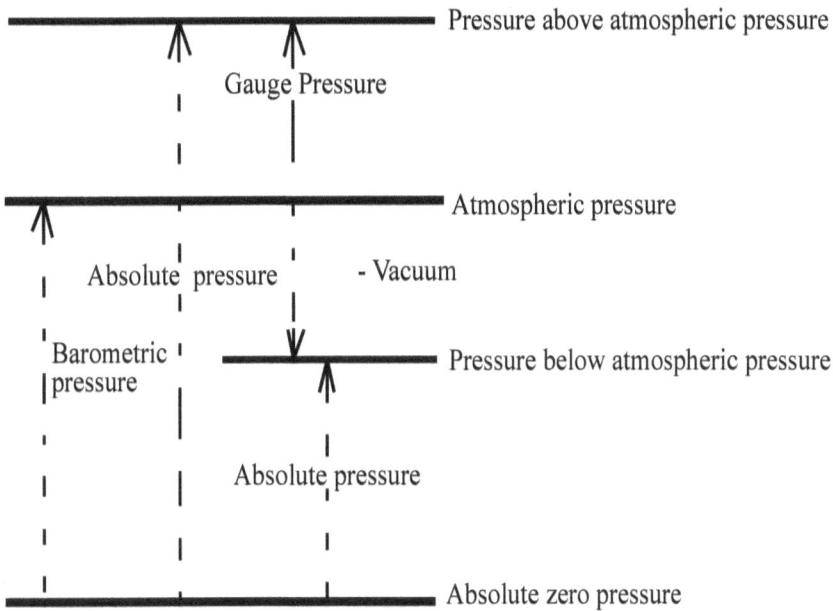

Fig. 6.1: Different types of pressure

Atmospheric pressure = Absolute pressure + Vacuum pressure

Vacuum pressure = - Gauge pressure.

The pressure measured above the perfect vacuum is called absolute pressure.

Principles of pressure measuring devices

1. By balancing the liquid column (whose pressure is to be found out) by the spring or dead weight. These are also called mechanical gauges. E.g. dead- weight testers, tyre pressure gauges.

2. By balancing the liquid column (whose pressure is to be found out) by the same or another column. These are also called as tube gauges. E.g. Piezometer tube, Manometer.

Pressure measuring devices

Diaphragm pressure gauge

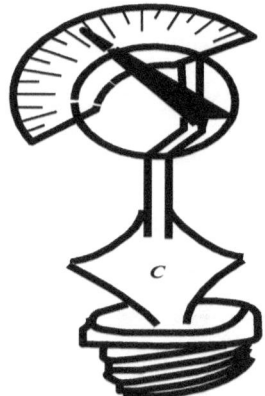

Fig. 6.2: Diaphragm pressure gauge

The spring side of the gauge is inserted in a closed chamber whose pressure is to be found out. The pressure is acting continuously on the spring. As the pressure is acting on the spring side, the deformation of spring or diaphragm occurs. The fluid pressure causes deformation of the diaphragm. The pointer rotates due to elastic deformation of the diaphragm which is due to rack and pinion arrangement. The pointer rotates over a calibrated scale which directly gives the pressure.

Piezometer tube

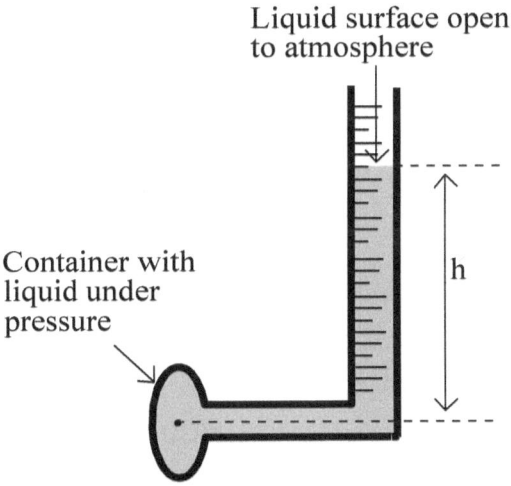

Fig. 6.3: Piezometer

It consists of a tube, one end of which is connected to the pipeline and the other end is opened to the atmosphere, in which the liquid can rise freely without overflow. The height to which the liquid rises gives the pressure head directly.

Precautions:

1. All burns and roughness near the hole must be removed, and the edge of the hole should be rounded off.

2. It is not suitable for measuring the negative pressure; as in such a case, the air will enter in the pipe through the tube.

Manometers

It is an improved form of piezometer tube which can measure comparatively high as well as negative pressures.

Based on the use and structure, following types of manometers are used:

i. Simple manometer

ii. Micro manometer

iii. Differential manometer

d. Inverted differential manometer

Simple manometer

A simple U- tube manometer is connected to the pipeline to determine pressure of the fluid. The liquid used in the bend tube or simple manometer is Mercury which is 13.6 times heavier than water. Hence, it is suitable for measuring high pressures.

The pipeline fluid exerts its pressure in the left limb. The displacement of the mercury in the left limb will cause a rise of the mercury level in the right limb as shown in Fig. 6.4 below.

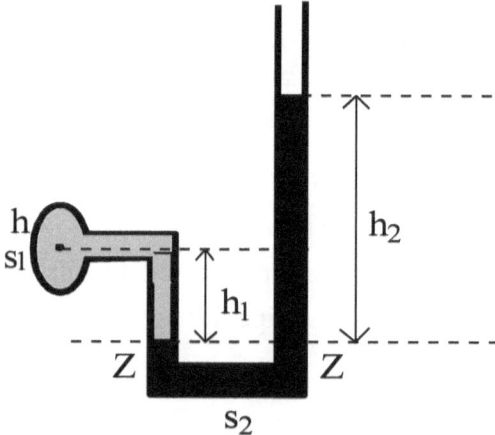

Fig. 6.4: A simple manometer to measure gauge pressure

84

The pressure in the left limb above Z-Z datum in the left limb;

$$= (h + s_1h_1) \text{ m of water} \dots\dots\dots (6.1)$$

The pressure in the right limb above Z-Z datum in the right limb;

$$= (s_2h_2) \text{ m of water} \dots\dots\dots (6.2)$$

Since, the pressure in both the limbs above Z-Z datum line are equal;

$$h + s_1h_1 = s_2h_2$$
$$\text{Or, } h = [s_2h_2 - s_1h_1] \text{ m of water} \dots\dots\dots (6.3)$$

Calculating negative pressure

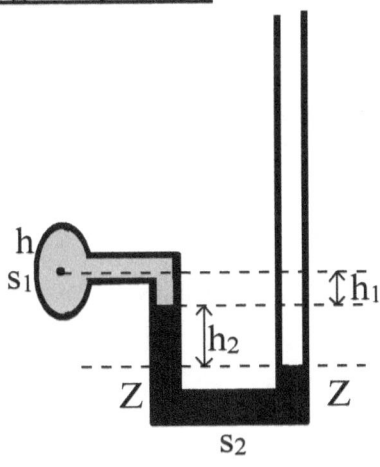

Fig. 6.5: A simple manometer to measure negative pressure

Pressure in the left limb above the datum line;

$$= [h + s_1h_1 + s_2h_2] \text{ m of water} \dots\dots\dots (6.4)$$

Pressure in the right limb above the datum line;

$$= 0 \dots\dots\dots (6.5)$$

Since, the pressure in both the limbs above Z-Z datum line are equal;

$$h + s_1h_1 + s_2h_2 = 0$$
$$\text{Or, } h = -[s_1h_1 + s_2h_2] \text{ m of water} \dots\dots\dots (6.6)$$

Example 6.1: Pressure in a simple manometer

1. A simple manometer containing mercury is used to measure the pressure of water flowing in a pipeline. The mercury in the open tube is 60mm higher than that on the left tube. If the height of the water in the left tube is 50mm, describe the pressure in the pipe in terms of head of water.

Sol^n:

Given,

Height of mercury in the left limb above Z-Z

(h_1) = 50mm = 0.05m

Height of mercury in the right limb above Z-Z (h_2) = 60mm = 0.06m

Let 'h' be the pressure in the pipe in terms of head of water.

s_1 and s_2 be the specific gravity of water and mercury respectively, such that s_1=1 and s_2=13.6.

Using equation 6.3, we have:

$h = [s_2h_2 - s_1h_1]$

$= [0.06 \times 13.6 - 1 \times 0.05] = 0.766$m of water.

Hence, the required pressure in the pipe in terms of head of water is 0.766m.

Exercises on simple manometer

1. A simple manometer containing mercury was used to find the negative pressure in the pipe as shown in the figure. The right limb of the manometer was opened to the atmosphere. Find the negative pressure, below the atmosphere in the pipe. [**Ans:** -7 m of water]

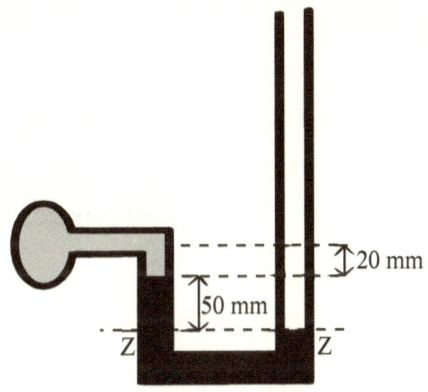

20 mm

50 mm

Z   Z

2. A simple manometer is used to measure the pressure of oil (sp. gr 0.8) flowing in a pipeline. Its right limb is opened to the atmosphere and the left limb is connected to the pipe. The centre of the pipe is 90 mm below the level of mercury in the right limb. If difference of mercury levels in the two limbs is 150 mm, find the pressure of oil in the pipeline. [Ans: 1.992 m of water]

3. A U-tube manometer is connected to a closed tank as shown in figure below. The air pressure in the tank is 120 Pa and the liquid in the tank is oil ($\gamma = 12000$ N/m^3). The pressure at point A is 20 kPa. Determine: (a) the depth of oil, z, and (b) the differential reading, h, on the manometer. [**Ans:** 103.2 kPa]

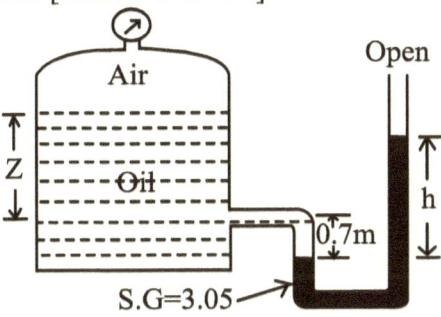

Open

Air

Z

Oil

0.7m

h

S.G=3.05

Micro manometer

It is a modified form of manometer in which the cross- sectional area of one of the limbs (say left limb) is made much larger (about 100 times) than that of the other limb for the measurement of low pressures where accuracy is of much importance.

Types of micro manometer:
a. Vertical tube manometer
b. Inclined tube manometer

a. Micro manometer [Vertical tube type]

The fluid flows in the left limb down to the basin. The movement of mercury in the basin will cause a considerable rise of the mercury in the right limb.

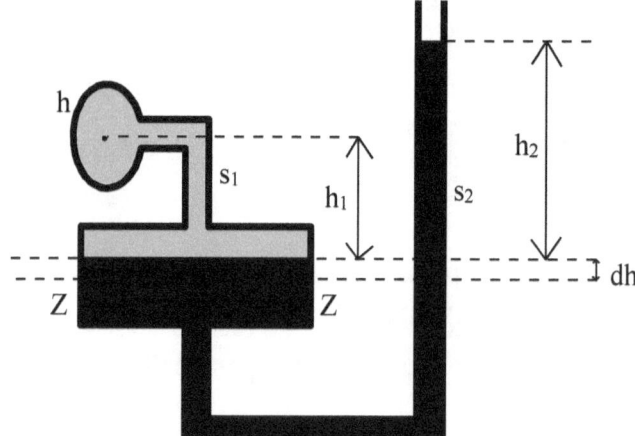

Fig. 6.6: Vertical tube type micro manometer

For a fall of a mercury level in the basin, there causes a corresponding rise of mercury in the right limb such that:

$$A.\delta h = a.h_2$$

$$\delta h = \frac{a}{A} \times h_2 \ldots\ldots\ldots (6.7)$$

Where a is the cross- sectional area of the tube in m^2.

Now, the pressure in the left limb above the datum line Z- Z is ;

$$= (h + s_1 h_1 + s_1 \delta h) \text{ m of water.......... (6.8)}$$

The pressure in the right limb above the datum line Z- Z is ;

$$= (s_2 h_2 + s_2 \delta h) \text{ m of water.......... (6.9)}$$

Since pressure in the both limbs above the Z-Z datum line are equal;

Hence from eq^n (6.8) and (6.9):

$$h = s_2 h_2 - s_1 h_1 + \delta h(s_2 - s_1) \text{ .... (6.10)}$$

Substituting the value of δh from eq^n (6.7) in eq^n (6.10);

$$h = s_2 h_2 - s_1 h_1 + \frac{a}{A} \times h_2 (s_2 - s_1) \text{.......... (6.11)}$$

If $A >>> a$; then $\dfrac{a}{A}$ can be neglected such that;

$$h = (s_2 h_2 - s_1 h_1) \text{ m of water.......... (6.12)}$$

Example 6.2: Pressure in a micromanometer

1. In order to determine the pressure in a pipe containing liquid of sp.gr. 0.8, a manometer was used as shown in the figure.

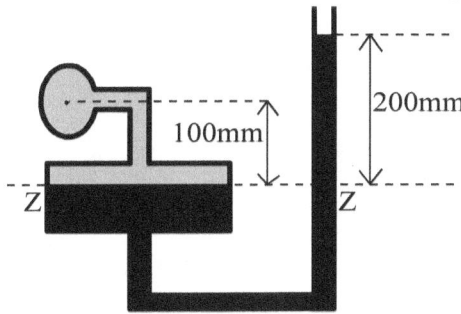

The ratio of area of the basin to that of the limb is 50. Find the pressure head in the pipe for the manometer reading.

Sol^n:

Given,

Height of mercury in the left limb above Z-Z (h_1) = 100mm = 0.1m

Height of mercury in the right limb above Z-Z (h_2) = 200mm = 0.2m

Specific gravity of pipeline fluid (s_1) = 08

Specific gravity of mercury (s_2) = 13.6

Let area of basin be 'A' and area of limb be 'a' such that

$$\frac{A}{a} = 50$$

Using equation 6.11, we have:

$$h = s_2h_2 - s_1h_1 + \frac{a}{A} \times h_2 \ (s_2-s_1)$$

$$= (13.6 \times 0.2) - (0.8 \times 0.1) + [50 \times 0.2(13.6-0.8)]$$

$$= 2.6912 \text{m of water.}$$

Hence, the pressure head in the pipe for the manometer reading is 2.6912m of water.

b. Micro manometer [Inclined tube type]

It is more sensitive than the vertical tube type.

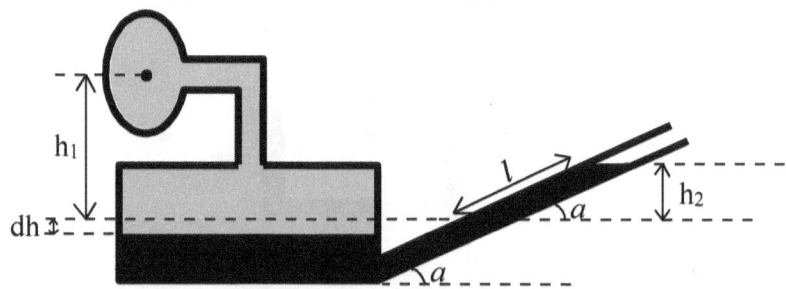

Fig. 6.7: Inclined tube type micro manometer

Due to inclination, the distance moved by the mercury in the narrow right limb will be comparatively more and hence, it gives a higher reading for the given pressure.

From geometry of the figure:

$Sin\alpha = h_2/l$

Or, $h_2 = l. Sin\alpha$

Substituting the value of h_2 in the final vertical micro manometer equation 6.11, we can find the required pressure in the pipeline.

Differential manometer

It is a device used for measuring the difference of pressures between two points in a pipe or in two different pipes. A differential manometer in its simplest form, consists of a U-tube containing a mercury whose two ends are connected to the points, whose difference in a pressure is to be found out.

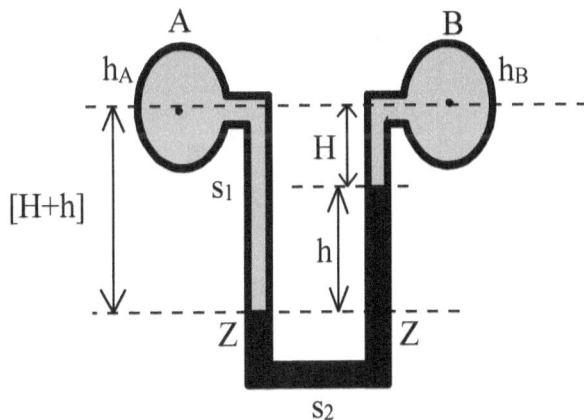

Fig. 6.8: Differential manometer

Consider a differential manometer connected at two points in a same pipeline containing a same fluid at same level.

The pressure head in the left limb above Z-Z datum line;

$= h_A + s_1 (H+h)$

$= (h_A + s_1H + s_1h)$ m of water .... (6.13)

The pressure head in the right limb above Z-Z datum line;

$= (h_B + s_1H + s_2h)$ m of water .... (6.14)

Since the pressure heads in both the limbs above Z-Z datum are equal, hence equating eq^n [6.13] & [6.14]:

$h_A + s_1H + s_1h = h_B + s_1H + s_2h$

Or, $h_A - h_B = s_2h - s_1h$

Or, $h_A - h_B = h(s_2 - s_1)$.......... (6.15)

Example 6.3: Pressure in a differential manometer

1. A differential manometer connected at the two points A and B at the same level in a pipe containing an oil of sp.gr. 0.8 shows a difference in mercury level as 100 mm. determine the difference in pressure at the two points. [**Ans: 1.280 m of water**]

Sol^n:

Given,

Difference in mercury level (h) = 100mm=0.1m

Specific gravity of oil $(s_1) = 0.8$

Specific gravity of mercury $(s_2) = 13.6$

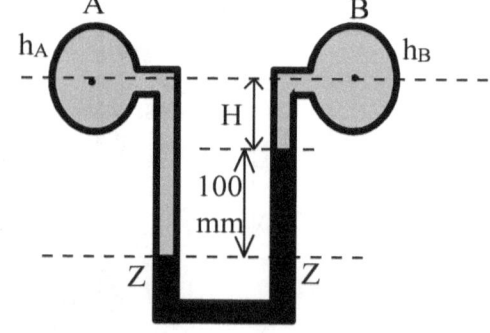

Let the pressure head in the pipe A and B be h_A and h_B.

Using equation 6.15, we have:

$h_A - h_B = h(s_2 - s_1) = 0.1 (13.6 - 0.8) = 1.280$m of water.

Hence, the required difference in pressure is 1.280m of water.

Exercises on differential manometer

1. A U-tube manometer filled with mercury is connected between two points in a pipeline. If the manometer reading is 26 mm of Hg, calculate the pressure difference between the points when water is flowing through the pipe. [**Ans:** 3214.4 N/m^2]

2. A U-tube manometer contains oil, mercury, and water as shown in Figure. For the column heights indicated what is the pressure differential between pipes A and B? [**Ans:** -20.7 KPa]

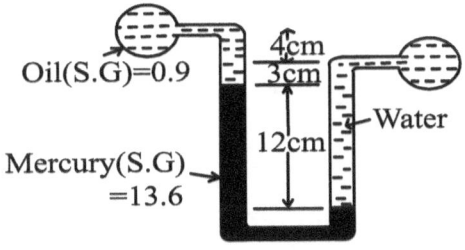

3. A U tube manometer measures the pressure difference between two points A and B in a liquid. The U tube contains mercury. Calculate the difference in pressure if h =1.5 m, h$_2$ = 0.75 m and h$_1$ = 0.5 m. The liquid at A and B is water (ω = 9.81 × 103N/m^2) and the specific gravity of mercury is 13.6. [**Ans:** 54.44 kN/m^2]

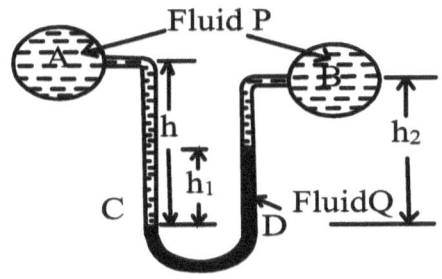

4. A U tube manometer measures the pressure difference between two points A and B in a liquid. The U tube contains mercury. Calculate the difference in pressure if h = 2.0 m, h_2 = 0.35 m and h_1 = 0.5 m. The liquid at A and B is oil (s = 0.85) and the specific gravity of mercury is 13.6.

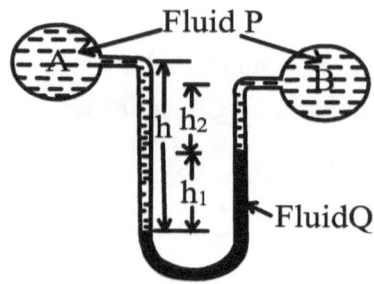

Inverted differential Manometer

It is used for measuring difference of low pressures, where accuracy is the prime consideration. It consists of an inverted U-tube, containing a light liquid whose two ends are connected to the points whose pressure difference is to be found out.

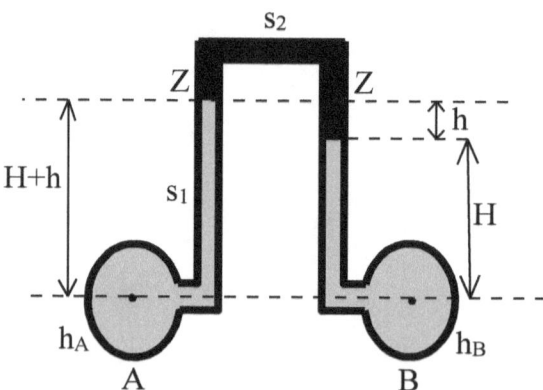

Fig. 6.9: Inverted differential manometer

The pipelines are at the same level with the same type of fluid.
The pressure head in the left limb below Z-Z datum line;

$= h_A - s_1 (H+h)$

$= (h_A - s_1 H - s_1 h)$ m of water .... (6.16)

The pressure head in the right limb below Z-Z datum line;

$= (h_B - s_1 H - s_2 h)$ m of water .... (6.17)

Since the pressure heads in both the limbs below Z-Z datum are equal, hence equating eq^n [6.16] & [6.17]:

$h_A - s_1 H - s_1 h = h_B - s_1 H - s_2 h$

Or, $h_A - h_B = -s_2 h + s_1 h$

Or, $h_A - h_B = h(s_1 - s_2)$.......... (6.18)

Example 6.4: Pressure in an inverted differential manometer

1. An inverted differential manometer having an oil of sp.gr 0.75 was connected to the two different pipes carrying water under pressure as shown in the figure.

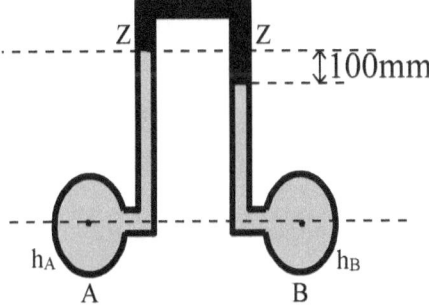

Determine the pressure in the pipe B in terms of kPa if the manometer reads as shown in the figure. Take pressure in the pipe A as 1.5m of water.

Solⁿ:

Given,

Pressure head in pipe A (h_A) = 1.5m

Specific gravity of water in pipeline (s_1) = 1

Specific gravity of oil (s_2) = 0.75

Difference in mercury level between two limbs (h) = 100mm= 0.1m

Pressure head in pipe B (h_B) =?

Using equation 6.18, we have:

$h_A-h_B = h(s_1 - s_2)$

Or, $1.5-h_B = 0.1(1-0.75)$

Or, $h_B = 1.475$m of water.

Therefore,

Pressure in pipe B = Pressure head in pipe B× density of water × gravity = $1.475 \times 1000 \times 9.8 = 14455$ Pa = 14.455 kPa

Hence, the pressure in the pipe B is found to be 14.455 kPa.

Exercises on inverted differential manometer

1. The inverted U-tube manometer of figure contains oil (SG = 0.9) and water as shown. The pressure differential between pipes A and B, $P_A - P_B$, is −5 kPa. Determine the differential reading, h. [**Ans: 0.46 mm**]

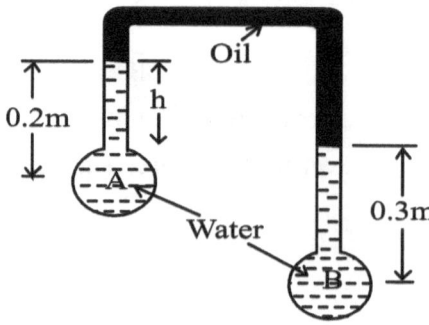

2. An inverted U tube as shown in the figure below is used to measure the pressure difference between two points A and B which has water flowing. The difference in level h = 0.3 m, a = 0.25 m and b = 0.15 m. Calculate the pressure difference $P_B - P_A$ if the top of the manometer is filled with:

(a) air

(b) oil of relative density 0.8.

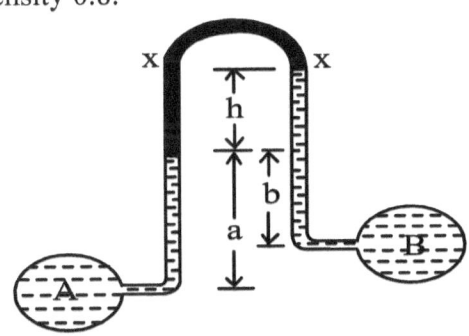

Chapter 7: Valves

Introduction

It is a mechanical device which regulates either the flow or the pressure of the fluid. Its function can be stopping or starting the flow, controlling the flow rate, diverting flow, preventing back flow, controlling pressure or relieving pressure.

Basically, the valve is an assembly of a body with connection to the pipe and some elements with a sealing functionality that are operated by an actuator. The valve can be also complemented with several devices such as positioners, transductors, pressure regulators, etc.

The three basic functions of valves are:

i. To stop flow

ii. To keep a constant direction of flow, and

iii. To regulate the flow rate and pressure.

Valves have many uses, including controlling water for irrigation, industrial uses for controlling processes, residential uses such as on / off and pressure control to dish and clothes washers and taps in the home. Even aerosols have a tiny valve built in. Valves are also used in the military and transport sectors.

Types of valves

On the basis of the operation of the valve closure member, valve can be classified as:

Multi- turn valve (linear motion valve)

The closure member has a linear displacement generally by turning its threaded stem several times. This operation is slow, which is necessary in some control valves. Eg: Gate valve, Globe valve, Fixed cone valve, Needle valve and Pinch valve.

Gate valve

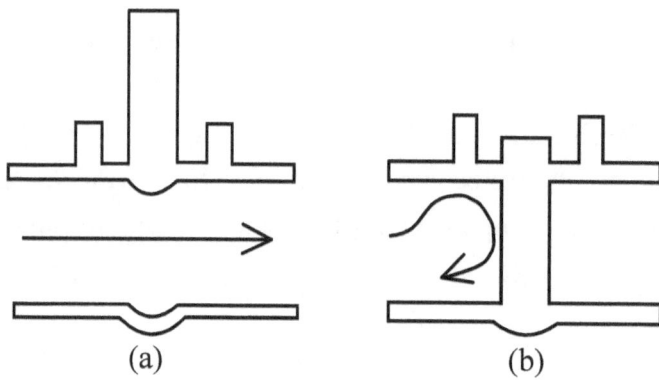

(a)   (b)

Fig. 7.1: Gate valve (a) Open (b) Closed

Like its name implies, the gate is lowered to cut off the path of flow. It is used as an on/off valve (not suitable as a control valve). There is little resistance to flow when fully open (allows smooth flow). However, long stroke requires time to open and close; not suitable for quick operation.

The distinct feature of a gate valve is the sealing surfaces between the gate and seats are planar, so gate valves are often used when a straight-line flow of fluid and minimum restriction is desired. The gate faces can be parallel, but are most commonly wedge-shaped. Gate valves are primarily used to permit or prevent the flow of liquids, but typical gate valves shouldn't be used for regulating flow, unless they are specifically designed for that purpose. Because of their ability to cut through liquids, gate valves are often used in the petroleum industry.

Gate valves are actuated by a threaded stem which connects the actuator (e.g. handwheel or motor) to the gate. They are

100

characterised as having either a rising or a nonrising stem, depending on which end of the stem is threaded. Rising stems are fixed to the gate and rise and lower together as the valve is operated, providing a visual indication of valve position. The actuator takes the form of a nut which is rotated around the threaded stem to move it. Nonrising stem valves are fixed to, and rotate with, the actuator, and are threaded into the gate. They may have a pointer threaded onto the upper end of the stem to indicate valve position, since the gate's motion is concealed inside the valve. Nonrising stems are used underground or where vertical space is limited.

Globe valve

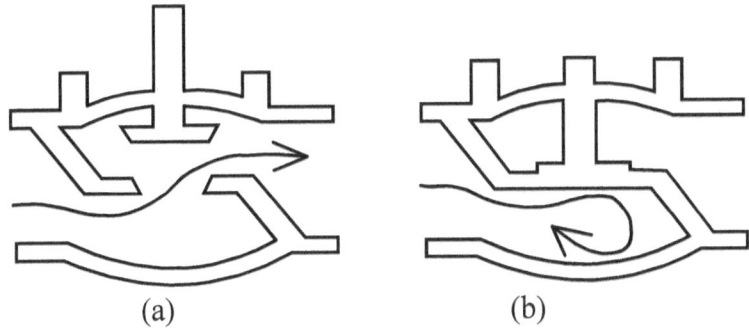

(a)   (b)

Fig. 7.2: Globe valve (a) Open (b) Closed

The globe-shaped body controls the fluid into an S- shaped flow. There is tight shut-off and can be used as a control valve. There is also large resistance to flow (does not allow smooth flow). Much power is required to open and close the valve (not suitable for large sizes).

A globe valve is a type of valve used for regulating flow in a pipeline, consisting of a movable disk-type element and a stationary ring seat in a generally spherical body. Globe valves are named for their spherical body shape with the two halves of

101

the body being separated by an internal baffle. This has an opening that forms a seat onto which a movable plug can be screwed in to close (or shut) the valve. The plug is also called a disc or disk. In globe valves, the plug is connected to a stem which is operated by screw action using a handwheel in manual valves. Typically, automated globe valves use smooth stems rather than threaded and are opened and closed by an actuator assembly.

Quarter- turn valve

The closure member as well as its shaft turn $0 - 90°$; from the fully open position to the fully- closed position. They are quick opening / closing valves. Eg: Ball valve, Butterfly valve, plug valve, spherical valve, etc.

Ball valve

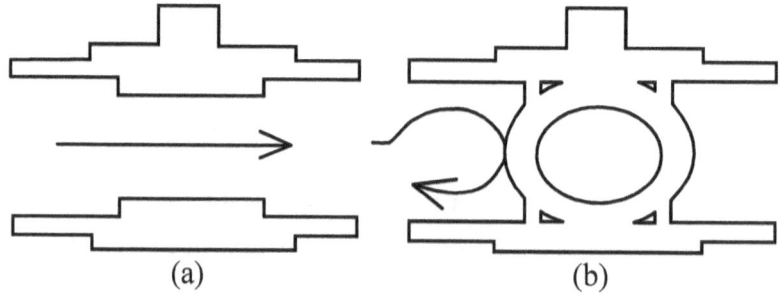

(a)   (b)

Fig.7.3: Ball valve (a) Open (b) Closed

Valve stopper is ball shaped. It is used as an on/off valve (not suitable as a control valve). There is little resistance to flow when fully open (allows smooth flow). Optimal for automated operation with a 90 degrees operating angle. However, advanced technology is required to manufacture ball.

102

The ball valve body design is conventional and flow is controlled by the position of a hole through the ball. This type of valve is used to control flow, pressure control in gas distribution systems and pressure reduction in connection with gas storage. These valves are also often used as isolating valves since they give rise to very little pressure drop in the fully open position.

A ball valve is a form of quarter-turn valve which uses a hollow, perforated and pivoting ball to control flow through it. It is open when the ball's hole is in line with the flow and closed when it is pivoted 90-degrees by the valve handle. The handle lies flat in alignment with the flow when open, and is perpendicular to it when closed, making for easy visual confirmation of the valve's status.

Butterfly valve

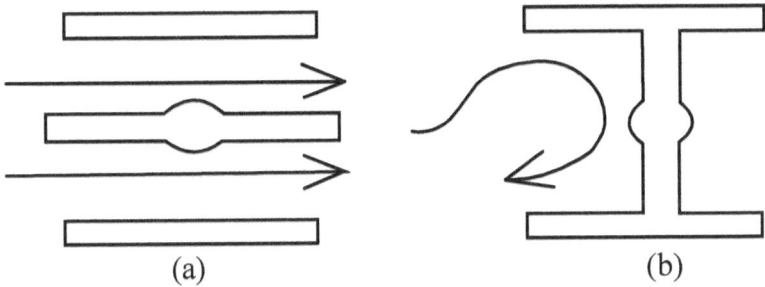

(a)   (b)

Fig. 7.4: Butterfly valve (a) Open (b) Closed

Valve shaped like a butterfly. There is a tight shut-off and can be used as a control valve. Little resistance to flow (allows smooth flow). Optimal for automated operation with a low operating torque and 90 degrees operating angle. Lightweight and compact (large diameter models are also available).

Operation is similar to that of a ball valve, which allows for quick shut off. Butterfly valves are generally favoured because they are lower in cost to other valve designs as well as being lighter in weight, meaning less support is required. The disc is positioned in the centre of the pipe; passing through the disc is a rod connected to an actuator on the outside of the valve. Rotating the actuator turns the disc either parallel or perpendicular to the flow. Unlike a ball valve, the disc is always present within the flow, so a pressure drop is always induced in the flow, regardless of valve position.

A butterfly valve is from a family of valves called quarter-turn valves. In operation, the valve is fully open or closed when the disc is rotated a quarter turn. The "butterfly" is a metal disc mounted on a rod. When the valve is closed, the disc is turned so that it completely blocks off the passageway. When the valve is fully open, the disc is rotated a quarter turn so that it allows an almost unrestricted passage of the fluid. The valve may also be opened incrementally to throttle flow.

Check valve

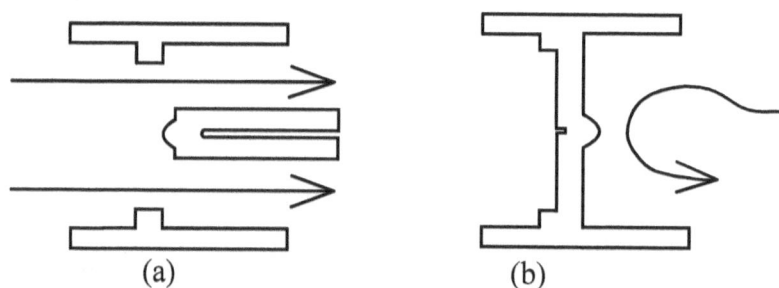

(a)   (b)

Fig. 7.5: Check valve (a) Open (b) Closed

They are used when the flow is unidirectional. It is for use when flow is only in one direction. The lightweight disc allows vertical installation. High operating speed prevents water hammer.

104

Check valves are two-port valves, meaning they have two openings in the body, one for fluid to enter and the other for fluid to leave. There are various types of check valves used in a wide variety of applications. Check valves are often part of common household items. Although they are available in a wide range of sizes and costs, check valves generally are very small, simple, or inexpensive. Check valves work automatically and most are not controlled by a person or any external control; accordingly, most do not have any valve handle or stem. The bodies (external shells) of most check valves are made of plastic or metal.

Chapter 8: Pumps

Introduction

A pump is a mechanical device to increase the pressure energy of a fluid. In most of the cases, pump is used to raising fluids from a lower level to a higher level. The raising of fluid from a lower to higher level is achieved by creating a low pressure at the inlet or suction and high pressure at the outlet or delivery end of the pump. Due to low inlet pressure, the fluid raises from a depth where it is available that the high outlet pressure forces it up to a height where it is required.

Pumps operate by some mechanism (typically reciprocating or rotary), and consume energy to perform mechanical work by moving the fluid. Pumps operate via many energy sources, including manual operation, electricity, engines, or wind power, come in many sizes, from microscopic for use in medical applications to large industrial pumps.

Mechanical pumps serve in a wide range of applications such as pumping water from wells, aquarium filtering, pond filtering and aeration, in the car industry for water-cooling and fuel injection, in the energy industry for pumping oil and natural gas or for operating cooling towers. In the medical industry, pumps are used for biochemical processes in developing and manufacturing medicine, and as artificial replacements for body parts, in particular the artificial heart and penile prosthesis.

Single stage pump - When in a casing only one impeller is revolving then it is called single stage pump.

<u>Double/ Multi stage pump</u> - When in a casing two or more than two impellers are revolving then it is called double/ multi stage pump.

Classification of pumps

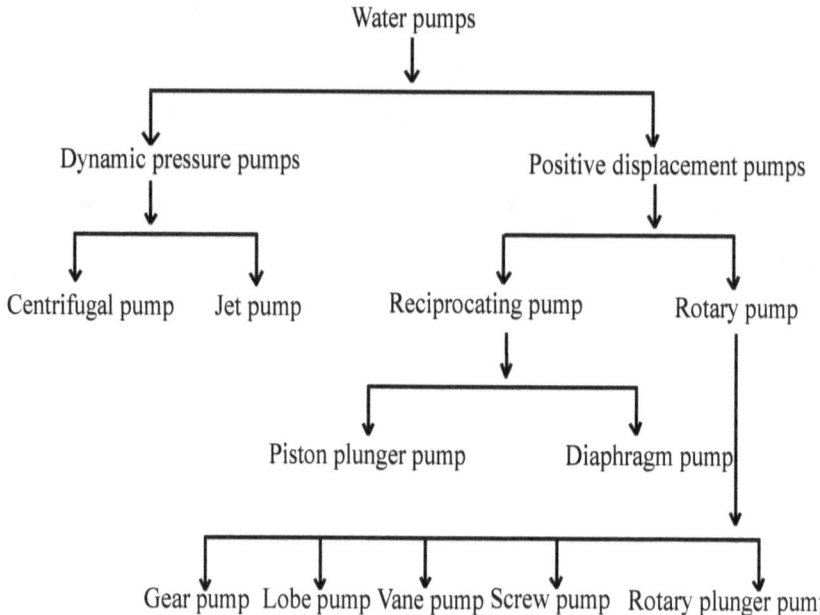

Fig. 8.1: Classification of hydraulic pumps

Mechanical pumps may be submerged in the fluid they are pumping or be placed external to the fluid. There are two basic types of pumps: positive displacement and centrifugal.

A positive displacement pump makes a fluid move by trapping a fixed amount and forcing (displacing) that trapped volume into the discharge pipe.

Some positive displacement pumps use an expanding cavity on the suction side and a decreasing cavity on the discharge side. Liquid flows into the pump as the cavity on the suction side expands

108

and the liquid flows out of the discharge as the cavity collapses. The volume is constant through each cycle of operation.

Centrifugal pump

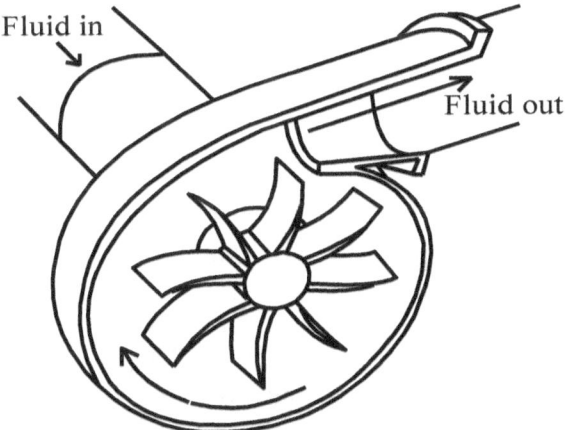

Fluid in

Fluid out

Fig. 8.2: Working of a centrifugal pump

A centrifugal pump is one of the simplest pieces of equipment in any process plant. Its purpose is to convert energy of a prime mover (an electric motor or turbine) first into velocity or kinetic energy and then into pressure energy of a fluid that is being pumped. The energy changes occur by virtue of two main parts of the pump, the impeller and the volute or diffuser. The impeller is the rotating part that converts driver energy into the kinetic energy. The volute or diffuser is the stationary part that converts the kinetic energy into pressure energy.

The process liquid enters the suction nozzle and then into eye (center) of a revolving device known as an impeller. When the impeller rotates, it spins the liquid sitting in the cavities between the vanes outward and provides centrifugal acceleration. As liquid leaves the eye of the impeller a low-pressure area is created causing more liquid to flow toward the inlet. Because the impeller blades are curved, the fluid is pushed in a tangential and radial direction by the centrifugal force. This force acting inside the pump is the same one that keeps water inside a bucket that is rotating at the end of a string.

The key idea is that the energy created by the centrifugal force is kinetic energy. The amount of energy given to the liquid is proportional to the velocity at the edge or vane tip of the impeller. The faster the impeller revolves or the bigger the impeller is, then the higher will be the velocity of the liquid at the vane tip and the greater the energy imparted to the liquid. This kinetic energy of a liquid coming out of an impeller is harnessed by creating a resistance to the flow. The first resistance is created by the pump volute (casing) that catches the liquid and slows it down. In the discharge nozzle, the liquid further decelerates and its velocity is converted to pressure according to Bernoulli's principle.

Multistage centrifugal pumps

A centrifugal pump containing two or more impellers is called a multistage centrifugal pump. The impellers may be mounted on the same shaft or on different shafts.

For higher pressures at the outlet, impellers can be connected in series. For higher flow output, impellers can be connected parallel.

A common application of the multistage centrifugal pump is the boiler feed water pump. For example, a 350 MW unit would require two feed pumps in parallel. Each feed pump is a multistage centrifugal pump producing 150 L/s at 21 MPa.

All energy transferred to the fluid is derived from the mechanical energy driving the impeller. This can be measured at isentropic compression, resulting in a slight temperature increase (in addition to the pressure increase).

Priming a pump

Typically, a liquid pump can't simply draw air. The feed line of the pump and the internal body surrounding the pumping mechanism must first be filled with the liquid that requires pumping: An operator must introduce liquid into the system to initiate the pumping. This is called priming the pump. Loss of prime is usually due to ingestion of air into the pump. The clearances and displacement ratios in pumps for liquids, whether thin or more viscous, usually cannot displace air due to its compressibility. This is the case with most velocity (rotodynamic) pumps — for example, centrifugal pumps.

Positive–displacement pumps, however, tend to have sufficiently tight sealing between the moving parts and the casing or housing of the pump that they can be described as self-priming. Such pumps can also serve as priming pumps, so called when they are used to fulfil that need for other pumps in lieu of action taken by a human operator.

Problems of centrifugal pumps

These are some difficulties faced in centrifugal pumps:

i. Cavitation- the net positive suction head (NPSH) of the system is too low for the selected pump

ii. Wear of the impeller- can be worsened by suspended solids

iii. Corrosion inside the pump caused by the fluid properties

iv. Overheating due to low flow

111

v. Leakage along rotating shaft

vi. Lack of prime - centrifugal pumps must be filled (with the fluid to be pumped) in order to operate

vii. A compressor stall - a local disruption of the airflow in a gas turbine or turbocharger compressor

Advantages

i. Simple in construction

ii. Low maintenance cost

iii. Low initial cost

iv. Quiet operation and requires less floor space.

Gear pump

Suction   Discharge

Fig. 8.3: Working of a gear pump

It consists essentially of two circular gears which mesh with each other. These run in close contact with the casing. The gears are rotated by some external agency. As spaces between the teeth of the impeller pass the suction opening, the liquid is caught between them, carried around the casing to the discharge opening. The teeth come into mesh, which forces the liquid out.

The external gear pump is a positive displacement (PD) type of pump generally used for the transfer and metering of liquids. The pump is so named because it has two gears that are side-by-side or external to each other. The gear pump is a precision machine with extremely tight fits and tolerances, and is capable of working against high differential pressures. The working principle of the external gear pump is illustrated in the figure. A drive gear (that is driven by a motor) rotates an idler gear in the opposite direction. When the gears rotate, the liquid, which is trapped in the gear teeth spaces between the housing bore and the outside of the gears, is transferred from the inlet side of the pump to the outlet side. It is important to note that the pumped liquid moves around the gears and not between the gears. The rotating gears continue to deliver a fresh supply of liquid from the suction (inlet) side of the pump to the discharge (outlet) side of the pump, with virtually no pulsations. The meshing of the gears on the discharge side of the pump forces the liquid out of the pump and into the discharge piping. In these pumps, the discharge rate is directly proportional to the speed.

They are generally used in:

i. Petrochemicals: Pure or filled bitumen, pitch, diesel oil, crude oil, lube oil etc.

ii. Chemicals: Sodium silicate, acids, plastics, mixed chemicals, isocyanates etc.

iii. Paint and ink.

iv. Resins and adhesives.

v. Pulp and paper: acid, soap, lye, black liquor, kaolin, lime, latex, sludge etc.

vi. Food: Chocolate, cacao butter, fillers, sugar, vegetable fats and oils, molasses, animal food etc.

Reciprocating pump

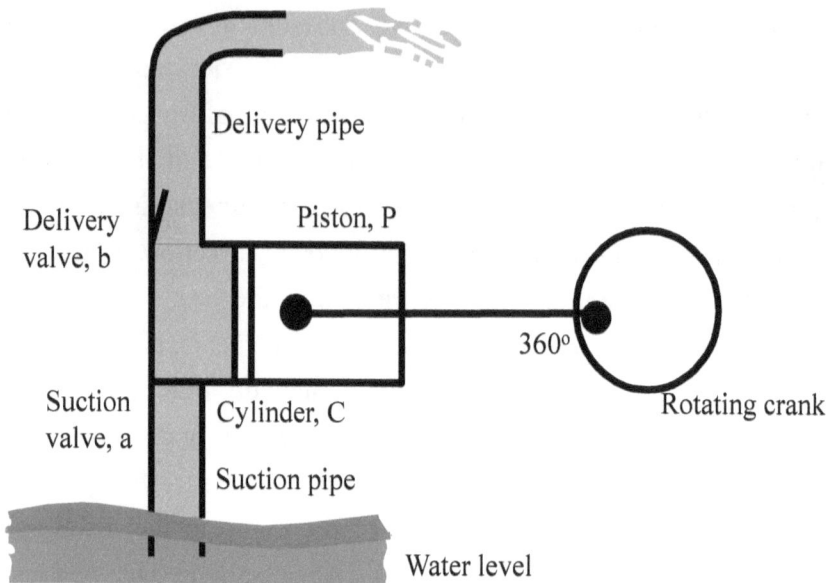

Fig. 8.4: Working of a reciprocating pump

A reciprocating pump is a class of positive-displacement pumps which is often used where a relatively small quantity of liquid is to be handled and where delivery pressure is quite large. In reciprocating pumps, the chamber in which the liquid is trapped, is a stationary cylinder that contains the piston or plunger.

Parts of a reciprocating pump

i. A cylinder in which Piston P works. The movement of the piston is obtained by a connecting rod, which connects the piston and the rotating crank.

ii. A suction pipe which connects the source of water and the cylinder.

iii. A delivery pipe into which the water is discharged from the cylinder.

iv. A suction valve "a" which admits the flow from the suction pipe into the cylinder.

v. A delivery valve "b" which admits the flow from the cylinder into the delivery pipe.

Working

Before starting the suction stroke, suction pipe and the clearance volume of the clearance are first filled with water to replace the air is known as the priming of the pump. Once the pump has been primed, water follows closely, the piston or the plunger on its forward strike. Movement of the piston or plunger creates a vacuum and the pressure forces the water up through the suction pipe into the cylinder.

This is the beginning of suction stroke (during this stroke, the piston P moves towards right from 0 to 180 degrees thus creating vacuum in the cylinder). This vacuum causes the suction valve "a" to open and the water enters the cylinder.

During the delivery stroke, the piston P, moves towards left, i.e. from 180 to 360 degree thus increasing pressure in the cylinder. This increase in pressure causes the suction valve 'a' to close and the delivery valve 'b' to open and the water is forced into the delivery pipe.

A reciprocating pump discharges a definite quantity of liquid during the displacement of its piston or plunger. This is why a reciprocating pump is ideally suitable for grouting operations in dam foundations.

It is best suited for relatively small capacities and high heads. In oil drilling operations, this type of pump is very common. Generally, it is used for light oil pumping, feeding of small boilers, pneumatic pressure systems.

Problems during pump operation

The problems that could occur when a pump is operating are:

i. Overloading: One risk of overloading is the danger of excess torque on a drive shaft.(May need a larger pump)

ii. Excess Speed: Running a pump at too high a speed causes loss of lubrication, which can cause early failure. Excess speed also runs a risk of damage from cavitation. (Use a higher displacement pump)

iii. Operating Problems: There are common operating problems in a pump.

   a. Pressure Loss: Pressure loss means that there is a high leakage path in a system.(relief valve, cylinders, motors, & A badly worn pump).

   b. Slow Operation: This can be caused by a worn pump or by a partial oil leak in a system. Pressure will not drop, however, if a load moves at all. Therefore, hp is still being used and is being converted into heat at a leakage point.

   c. No Delivery: If oil is not being pumped, a pump-

   1. Could be assembled incorrectly.
   2. Could be driven in the wrong direction.
   3. Has not been primed. The reasons for no prime are usually improper start-up, inlet restrictions, or low oil level in a reservoir.
   4. Has a broken drive shaft.

116

iv.Noise: If you hear any unusual noise, shut down a pump immediately. Cavitation noise is caused by a restriction in an inlet line, a dirty inlet filter, or too high a drive speed. Air in a system also causes noise. Noise can be caused by worn or damaged parts, which will spread harmful particles through a system, causing more damage if an operation continues.

v. Cavitation: Cavitation occurs where available fluid does not fill an existing space.

Most of the time cavitation occurs in the suction part of the system. When cavitation takes place the pressure in the fluid decreases to a level below the ambient pressure thus forming 'vacuumholes' in the fluid.

When the pressure increases, for example in the pump, these 'vacuumholes' implode.

Cavitation can be caused by:

a. acceleration of the oil flow behind a throttle when the oil contains water or air
b. high fluid temperature
c. a resistance in the suction part of the system
d. a suction line which is to small in diameter
e. a suction hose with a damaged inside liner
f. a suction filter which is saturated with dirt (animation)
g. high oil viscosity
h. insufficient breezing of the reservoir

Alternating current

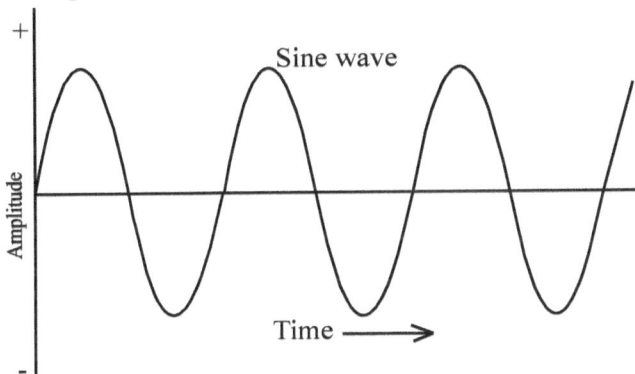

Fig. 9.1: Alternating current

In AC, the flow of electric charge changes periodically and reverses direction. The current or voltage alternates its direction and magnitude every time. Nowadays 95% of the total energy is produced, transmitted and distributed in AC supply.

Merits of AC over DC

i. More voltage can be generated than DC (upto 33000 volts when DC only 650 volts).

ii. AC voltage can be increased or decreased by transformer.

iii. AC transmission and distribution is more economical as more power can be transmitted at higher voltage.

iv. AC motors for the same Horsepower as of DC motors are cheaper, lighter in weight, requires less space and requires lesser attention in operation and maintenance

v. AC can be converted to DC easily when and where required but DC cannot be converted to AC so easily and it will not be economical.

Direct current

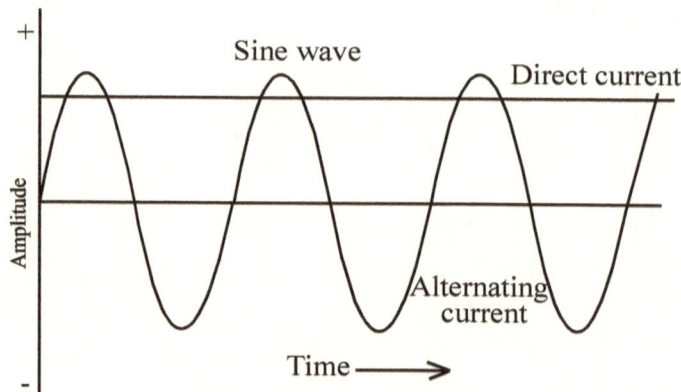

Fig. 9.2: Direct current

DC is the unidirectional flow of electric charge. Its direction and magnitude is steady at all the times. DC is produced by sources like batteries, thermocouples, solar cells, dynamos or generators.

Merits of DC over AC

i. DC series motor are most suitable for traction purposes, tramway, railway, trains and lifts.

ii. For electrotyping, electroplating and electrochemical purposes.

iii. AC lamps for search lights and cinema projectors work on DC.

iv. Arc welding is better than on AC.

v. Relay and operating time switches and circuit breakers works efficiently on DC.

vi. In rolling mills, paper mills where the fine speed control is required in both the directions, DC motors are required.

Single phase current

The single phase electric power refers to the distribution of AC electric power using a system in which all the voltages of the supply

120

vary in unison. Single phased electric current is a cycle of voltage that operates in the same time phase as shown in the figure 9.3.

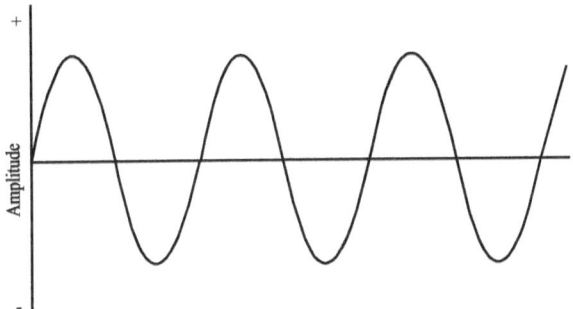

Fig. 9.3: Single phase alternating current

Single phase power distribution is widely used especially in rural areas where the cost of 3 phase distribution network is high and the motor loads are small and uncommon.

Three phase current

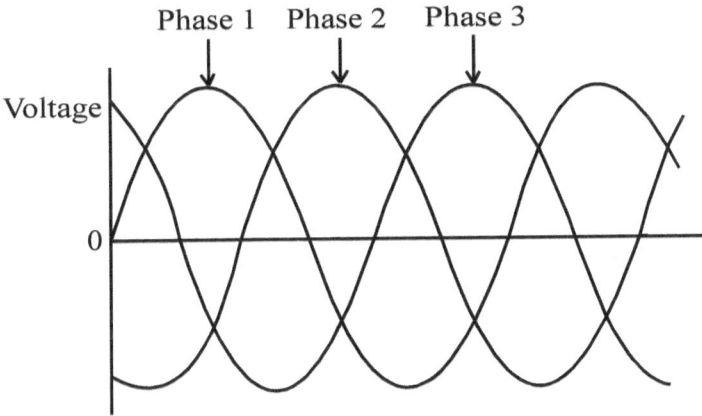

Fig. 9.4: Triple phase alternating current

In 3 phase system, the current or voltage reaches their peak in instantaneous value sequentially not simultaneously in each cycle. Firstly, the first phase, then the second, then the third reaches its maximum values. In 3- phase system, we have 3 distinct source of power in 3 distinct time zones as shown in Fig. 9.4. The waveform of the three supplied semiconductors are offset from one another in time by one-third of their period. Higher power systems i.e. 100 of KVA or larger are nearly always three phase.

Power factor (P.f)

The power factor of an AC electrical power system is defined as the ratio of the real power flowed to the apparent power of the load in the circuit. It is dimensionless number ranging from 0 to 1.

Real power is the capacity of the circuit for performing work in a particular time. Apparent power is the product of current and voltage in the circuit.

$$P.f = \frac{Real\ power}{Apparent\ power} = \frac{V \times I \times Cos\emptyset}{(V \times I)} \quad \cdots\cdots\cdots\ (9.1)$$

where, P.f is the power factor, V is the supplied voltage, I is the current flowing in the circuit and $Cos\emptyset$ is the cosine of the angle of load or lag of the current from applied voltage.

P.f improvement is necessary in an installation because low p.f has the following adverse effect:

i. Low p.f draws more current increasing wasting and decreases efficiency.

ii. With low p.f generators, transformers, switches, transmission lines become overloaded and increases voltage.

iii. With low p.f. , cost of generation and transmission increases due to the increase in current and the use of thicker wires and bigger switches.

122

The linear loads with low p.f can be corrected with a passive network of capacitors or inductors.

Transformer

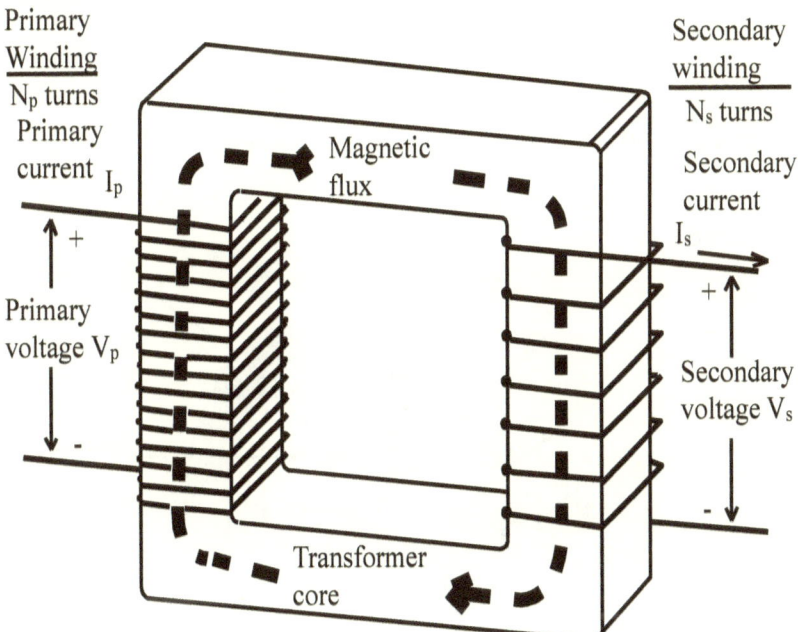

Fig. 9.5: Primary and secondary winding of a transformer

The transformer may be defined as the static piece of electrical apparatus which converts electrical power from one circuit to other circuit at the same frequency. It can increase or decrease voltage with corresponding decrease or increase of current keeping the power same. This transformation of energy is done due to the Faraday's law of electromagnetic induction through the 2 windings-primary and secondary as shown in Fig. 9.5.

The transformer moves on the principles of mutual induction action i.e. if the two coils are placed near to each other and if one is

123

connected to the AC supply, the emf will be induced in the other coil. This emf induced will be according to the number of turns in the secondary coil.

$$\frac{Vp}{Vs} = \frac{Np}{Ns} \quad\text{..........} \quad (9.2)$$

The transformer is mostly used in AC supply to increase the voltage and hence decrease the current and keep the same power for transmitting high voltage or to decrease the voltage at the load point.

AC motor

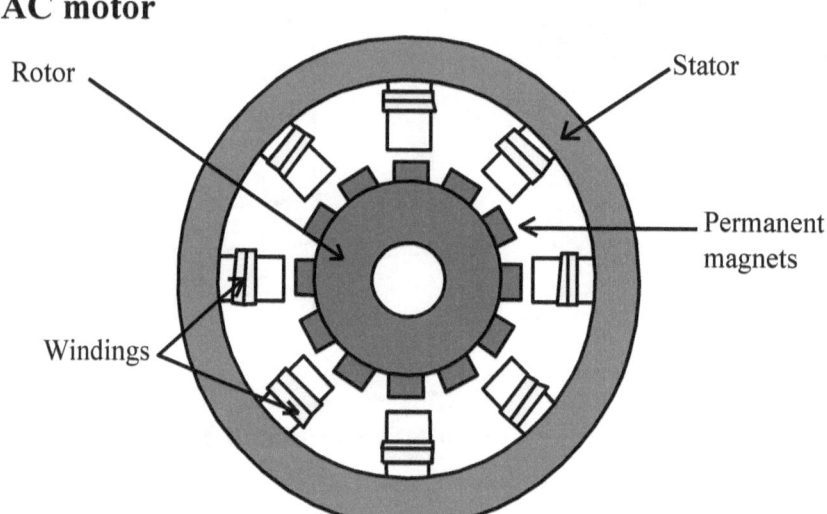

Fig. 9.6: Basic AC Motor components

It is an electrochemical device that converts electrical energy into mechanical energy. The interaction between an electric motor, magnetic field and the winding current generates spores and is useful on the motor. Eg: Industrial fans, blowers, pump lines, compressors, etc. Devices such as magnetic solenoids and transducers convert electricity into motion but do not generate usable mechanical power. Hence they are also referred to as actuators.

An AC motor has 2 parts:

a. The stationery stator having coils supplied with AC to produce rotating magnetic field.

b. A rotor attached to the output shaft i.e. given by torque by the rotating field.

When the stator of a single phase motor is fed with single phase supply, it produces alternating flux in the stator winding. The alternating current flowing through stator winding causes induced current in the rotor bars according to Faraday's law of electromagnetic induction. This induced current in the rotor will also produce alternating flux. If the rotor is given a initial start by external force in either direction, then motor accelerates to its final speed and keeps running with its rated speed.

Generator

An electrical generator is a device that converts mechanical energy into electrical energy. A generator forces electrical energy (usually carried by electrons) to flow through an external electrical circuit. The source of mechanical energy may be reciprocating or turbine steam energy, water falling through a turbine or water wheel, an internal combustion engine, a wind turbine, a hand crank, compressed air or any other sources of mechanical energy. Eg: Alternator, Dynamo, Inductor, Generator, etc.

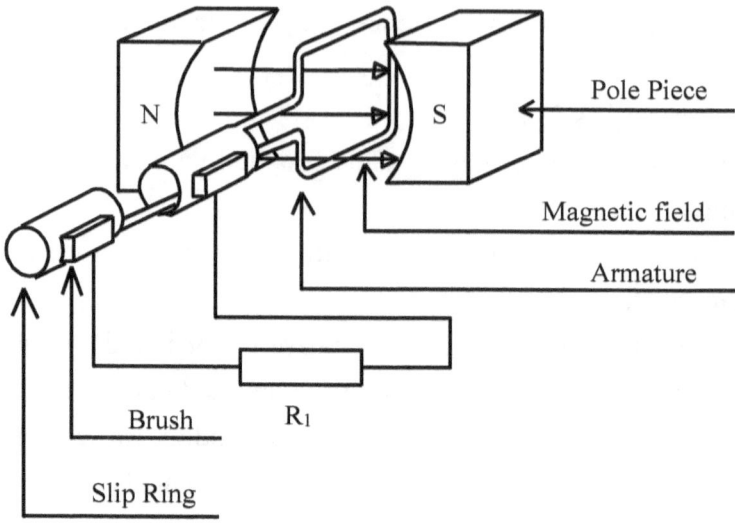

Fig. 9.7: AC generator

AC generator has the following salient parts:

i. Rotor: The rotor is a non- stationary part which rotates because the wire and the magnetic field of the motor are arranged so that the torque is developed about the rotor axis.

ii. Stator: The stator is stationary part of the rotary system. It consists of steel frame enclosing a hollow cylindrical core made up of laminations of silicon steel to reduce hysteresis and eddy current productions.

iii. Excitor: The excitor is generally a DC shunt or a compound generator which produces voltage upto 250 volt.

Fuses and Switches

Fuse is a type of low resistance resistor device that prevents over current production, short circuit, mismatched loads or device failure are the prime reasons of excessive current. Fuse consists of fused elements (Zn, Cu, Ag, Al or alloy) whose metal wire or strip melts

126

when too much current flows and intercepts the circuit in which it is connected.

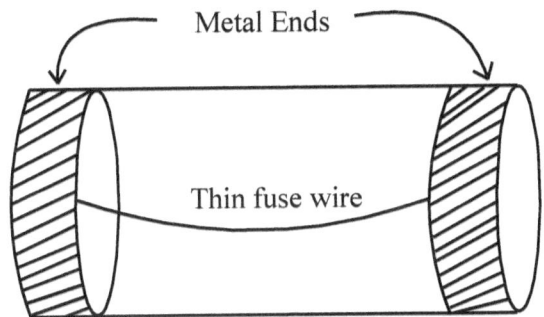

Fig. 9.8: Electrical fuse

The fuse interrupts excessive current so that further damage by overheating or fire is prevented.

A switch is an electrical component that can break an electrical circuit interrupting the current or diverting it from one conductor to another.

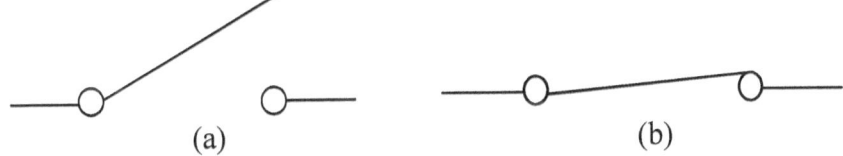

Fig. 9.9: Circuit symbol for (a) Open switch (b) Closed switch

Chapter 10: Mechanical power transmission

Introduction

Power transmission is the movement of energy from its place of generation to a location where it is applied to performing useful work. Mechanical power may be transmitted directly using a solid structure such as driveshaft, transmission gears can adjust the amount of torque or force vs speed in much same the way as an electrical transformer adjusts voltage vs current.

Hydraulic systems uses liquid under pressure to transmit power; canals and hydroelectric power generation facilities harness natural water power to lift ships or generate electricity.

Pneumatic systems uses gases under pressure to transmit power; compressed air is commonly used to operate pneumatic tools in factories and repair gauges.

Mechanical power transmission includes chains, sprockets, belts, pulleys, gears, clutches, etc.

Belt

A belt is a loop of flexible material (leather, cotton, woven fabrics, silk, rubber, etc) used to mechanically link two or more rotating shafts to transmit power efficiently. Belts are looped over pulleys.

The amount of power transmission depends upon:

i. The velocity of the belt
ii. The tension under which the belt is placed over pulleys
iii. The arc of contact between the belt and the smaller pulley
iv. The conditions under which the belt is used.

Types of belt

There are 4 types of belt: Flat, V- shaped, toothed or timing and round belt.

i. Flat belt

It is used in factories and workshop where only fraction of power is to be transmitted.

It is used when the distance between the two pulleys are short (< 8m).

ii. V- belts

It is used where a great amount of power is to be transmitted between the 2 pulleys.

iii. Circular belt/ rope

It is used to transmit great amount of power between 2 pulleys (> 8m).

iv. Timing belt

These have teeth that fit into the matching pulley. Timing belts need the least tension of all belts and are among the most efficient.

Types of flat belt drives

Open belt drive

The open belt drive, as shown in Fig. 10.1, is used with shafts arranged parallel and rotating in the same direction. In this case, the driver *A* pulls the belt from one side (*i.e.* lower side *RQ*) and delivers it to the other side (*i.e.* upper side *LM*).

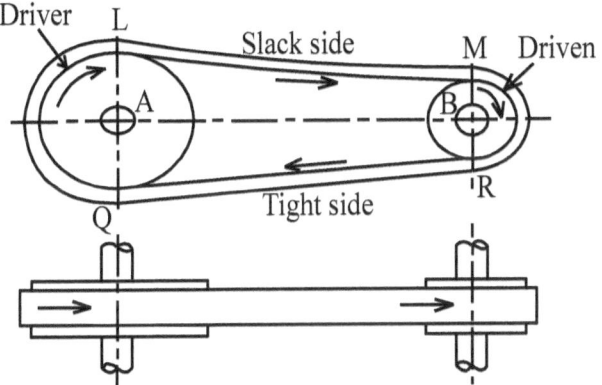

Fig.10.1: Open belt drive

Thus the tension in the lower side belt will be more than that in the upper side belt. The lower side belt (because of more tension) is known as *tight side* whereas the upper side belt (because of less tension) is known as *slack side*, as shown in Fig. 10.1.

Crossed or twist belt drive

It is used when the shafts are arranged parallel and rotating in opposite directions. In this case, the driver pulls the belt from one

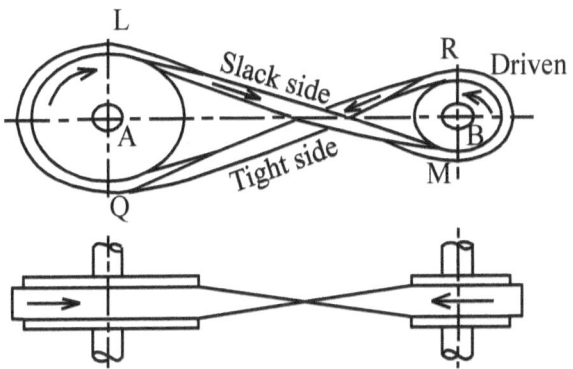

Fig. 10.2: Crossed belt drive

side (*i.e. RQ*) and delivers it to the other side (*i.e. LM*). Thus the tension in the belt *RQ* will be more than that in the belt *LM*. The belt *RQ* (because of more tension) is known as **tight side,** whereas the belt *LM* (because of less tension) is known as **slack side,** as shown in Fig.10.2. A little consideration will show that at a point where the belt crosses, it rubs against each other and there will be excessive wear and tear. In order to avoid this, the shafts should be placed at a maximum distance of 20 *b*, where *b* is the width of belt and the speed of the belt should be less than 15 m/s.

Quarter- turn belt drive

The quarter turn belt drive also known as right angle belt drive, as shown in the Fig. 10.3, is used with shafts arranged at right angles and rotating in one definite direction.

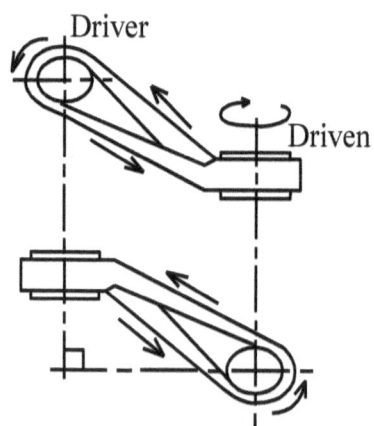

Fig. 10.3: Quarter- turn belt drive

In order to prevent the belt from leaving the pulley, the width of the face of the pulley should be greater or equal to 1.4 b, where b is the width of belt.

Belt drive with idler pulley

A belt drive with an idler pulley, as shown in Fig. 10.4 is used with shafts arranged parallel and when an open belt drive cannot be used due to small angle of contact on the smaller pulley.

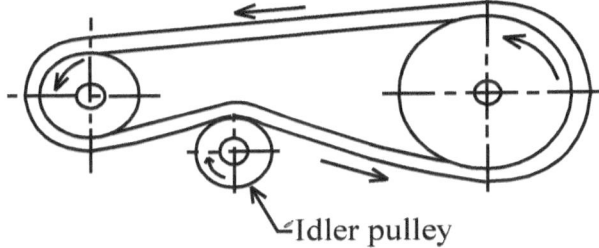

Fig. 10.4: **Open belt drive with idler pulley**

This type of drive is provided to obtain high velocity ratio and when the required belt tension cannot be obtained by other means.

Compound belt drive

A compound belt drive, as shown in the Fig. 10.5, is used when power is transmitted from one shaft to another through a number of pulleys.

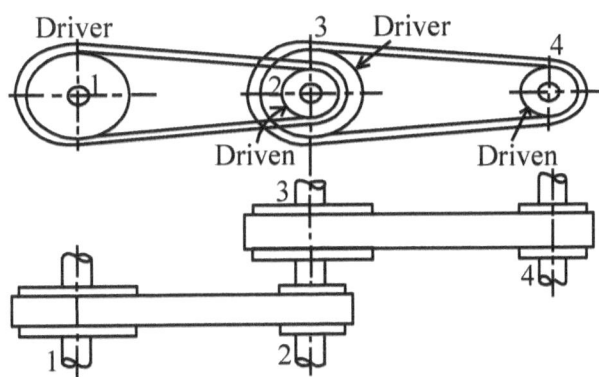

Fig. 10.5: **Compound belt drive**

It is made up of two or more compound doubled-flanged wheels linked by means of tense belts. The performance of this

133

mechanism is similar to that of a simple belt drive. Every compound double-flanged wheel couples two or more single wheels to the same axis or shaft.

Stepped or cone pulley drive

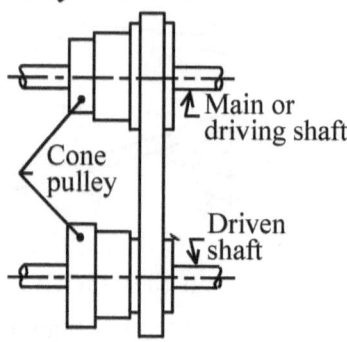

Fig. 10.6: Stepped or cone pulley drive

A stepped or cone pulley drive, as shown in Fig. 10.6, is used for changing the speed of the driven shaft while the main or driving shaft runs at constant speed. This is accomplished by shifting the belt from one part of the steps to the other.

Fast and loose pulley drive

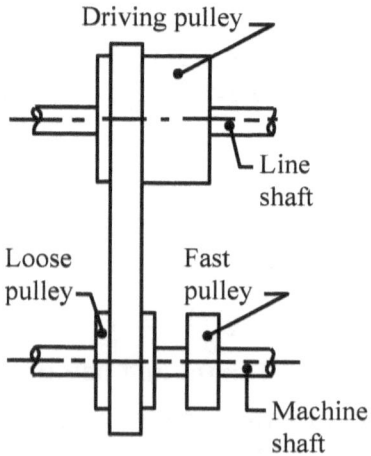

Fig. 10.7: Fast and loose pulley drive

A fast and loose pulley drive, as shown in Fig. 10.7, is used when the driven or machine shaft is to be started or stopped whenever desired without interfering with the driving shaft.

A pulley which is keyed to the machine shaft is called *fast pulley* and runs at the same speed as that of machine shaft. A loose pulley runs freely over the machine shaft and is incapable of transmitting any power. When the driven shaft is required to be stopped, the belt is pushed on to the loose pulley by means of sliding bar having belt forks.

Velocity ratio of belt drives

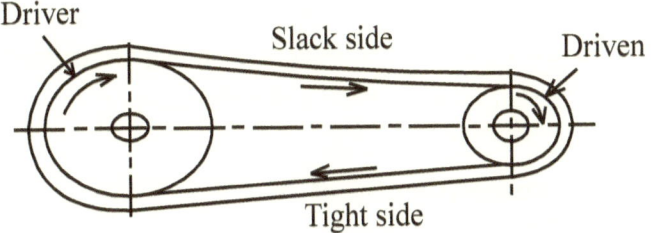

Fig. 10.8:Cross-section of a flat belt

It is the ratio between the velocities of the driver and the follower or driven.

Let, d_1 = diameter of the driver, d_2 = diameter of the driven, N_1 = speed of the driver in rpm and N_2 = speed of the driven in rpm.

The length of the belt that passes over the driver in one minute

= $\pi d_1 N_1$ .... (10.1)

Similarly, the length of the belt that passes over the driven in one minute = $\pi d_2 N_2$ .... (10.2)

Since the length of the belt that passes over the driver in one minute is equal to the length of the belt that passes over the follower in one minute;

$$\pi d_1 N_1 = \pi d_2 N_2$$

Or, $\dfrac{N_2}{N_1} = \dfrac{d_1}{d_2} = $ Velocity ratio.......... (10.3)

When the thickness of the belt 't' is considered;

$$\dfrac{N_2}{N_1} = \dfrac{(d_1 + t)}{(d_2 + t)} \quad .... (10.4)$$

For compound belt drive:

$$\dfrac{N_2}{N_1} \times \dfrac{N_4}{N_3} = \dfrac{d_1}{d_2} \times \dfrac{d_3}{d_4} \quad .... (10.5)$$

Or, $\dfrac{N_4}{N_1} = \dfrac{d_1 \times d_3}{d_2 \times d_4}$ as, $[N_2 = N_3]$.......... (10.6)

Hence,

$$\dfrac{\text{speed of last driven}}{\text{speed of first driver}} = \dfrac{\text{Product of diameters of drivers}}{\text{Product of diameter of drivens}}$$

Exercises on velocity ratio of belt drive

1. In a crossed belt drive system of power transmission between 2 parallel shafts, find the diameter of the driven shaft pulley which is required to run at 125 rpm by the driving shaft pulley diameter 500mm and speed of 25 rpm. [**Ans:** 100 mm]

2. It is required to drive a shaft at 680 rpm by means of a belt from a parallel shaft A having a pulley 300mm diameter on it and running at 340 rpm. What sized pulley is required on the shaft B? [**Ans:** 150 mm]

Power transmitted by a belt

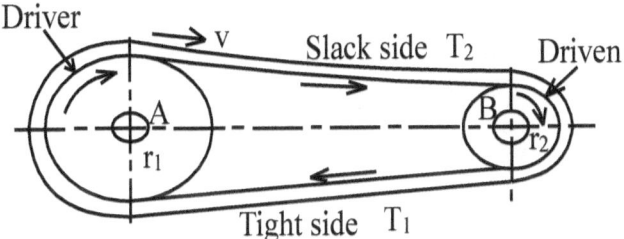

Fig. 10.9: Power transmitted by open belt drive

Consider a driving pulley A and the driven pulley B with greater tension in the former side than the latter as shown in the figure.

Let, T_1 and T_2 be the tensions in the tight and slack side of the belt respectively in Newtons.

r_1 and r_2 be the ratio of the driver and follower respectively

And, v = velocity of the belt in m/s

The effective driving force at the circumference of the follower is the difference between the two tensions $(T_1 - T_2)$.

Therefore, Workdone per second = $(T_1 - T_2).v$ Nm/s

And, Power transmitted (P) = $(T_1 - T_2)$ v Watt.......... **(10.7)**

Slip of the belt

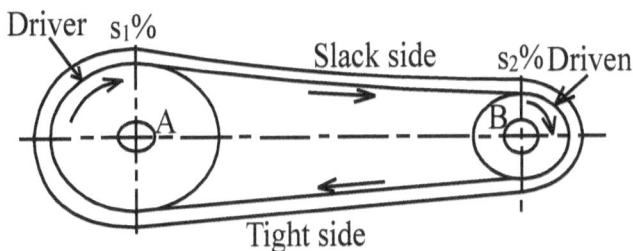

Fig. 10.10: Slip of open belt drive

When the frictional grip becomes insufficient between the belts and the shafts, the driver moves with the forward motion without carrying the belt with it which is called the slip of the belt and generally expressed as percentage.

Let, $S_1\%$ = slip between the driver and the belt, and

$S_2\%$ = slip between the belt and the follower

d_1 = diameter of the driver, d_2 = diameter of the driven, N_1 = speed of the driver in rpm and N_2 = speed of the driven in rpm.

Then,

The velocity of the belt passing over the driver per second;

$$v = [\frac{\pi d_1 N_1}{60}] - [\frac{\pi d_1 N_1}{60}] \times s_1\% = \frac{\pi d_1 N_1}{60} [1 - (\frac{s_1}{100})] \dots \dots (10.8)$$

The velocity of the belt passing over the follower per second;

$$\frac{\pi d_2 N_2}{60} = v - v s_2\% = v [1 - (\frac{s_2}{100})] \dots \dots (10.9)$$

From equation [11.8] and [11.9]:

$$\frac{\pi d_2 N_2}{60} = \frac{\pi d_1 N_1}{60} [1 - (\frac{s_1}{100})] [1 - (\frac{s_2}{100})]$$

$$Or, \frac{N_2}{N_1} = \frac{d_1}{d_2} [1 - (\frac{s_1}{100}) - (\frac{s_2}{100})] \quad [neglecting (\frac{s_1 s_2}{100 \times 100})]$$

$$= \frac{d_1}{d_2} [1 - \frac{(s_1 + s_2)}{100}]$$

Therefore, $\frac{N_2}{N_1} = \frac{d_1}{d_2} [1 - (\frac{s}{100})] \dots \dots (10.10)$

$[s = s_1 + s_2 = $ total % of slip$]$

For compound belt drive:

$$\frac{N_2}{N_1} \times \frac{N_4}{N_3} = \frac{d_1}{d_2} \times \frac{d_3}{d_4} [1 - (\frac{s}{100})] \dots \dots (10.11)$$

$$Or, \frac{N_4}{N_1} = \frac{d_1 \times d_3}{d_2 \times d_4} [1 - (\frac{s}{100})] \dots \dots (10.12)$$

138

Example 10.1: Slip of the belt drive

1. An engine running at 350 rpm drives a line shaft by means of a belt. The engine pulley is 750 mm diameter and the pulley on the line shaft is 450mm. A 900 diameter pulley on the line shaft drives a 250mm diameter keyed to a dynamo shaft. Find the speed of the dynamo shaft when: i. there is no slip, and ii. There is total slip of 4%.

Sol^n:

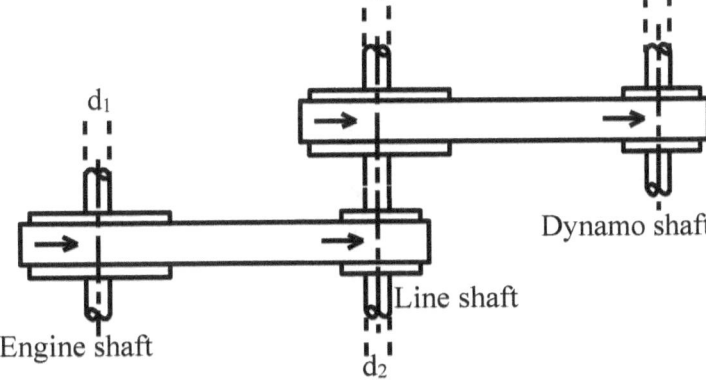

Given,

Speed of the engine shaft (N_1) = 350 rpm

Diameter of engine shaft pulley (d_1) = 750mm

Diameter of line shaft pulley (d_2 or d_3) = 450mm

Diameter of dynamo shaft pulley (d_4) = 250mm

Total slip% (s) = 4

Speed of dynamo shaft (N_4) =?

i. When there is no slip, using equation 10.6:

$$\frac{N_4}{N_1} = \frac{d_1 \times d_3}{d_2 \times d_4}; \text{ Or, } N_4 = \frac{750 \times 900 \times 350}{450 \times 250} = 2100 \text{ rpm.}$$

ii. When there is slip of 4%, using equation 10.12:

$$\frac{N_4}{N_1} = \frac{d_1 \times d_3}{d_2 \times d_4}[1-(\frac{s}{100})]$$

139

Or, $N_4 = \dfrac{750 \times 900 \times 350}{450 \times 250}\left[1-\left(\dfrac{4}{100}\right)\right] = 2017$ rpm.

Hence, the speed of the dynamo shaft is 2100 rpm and 2017 rpm with no slip and 4% slip respectively.

Exercises on slip of the belt

1. A shaft running at 200 rpm is to drive a parallel shaft at 300 rpm. The pulley on the driving shaft is 60 cm diameter. Calculate the diameter of the pulley on the driven shaft when: a. there is no slip b. there is a total slip of 4%. [**Ans:** 40cm, 39.83 cm and 38.22 cm]

2. With the help of a belt, an engine running at 200 rpm drives a line shaft. The diameter of the pulley on the engine is 80 cm and the diameter of the pulley on the line shaft is 40 cm. A 100 cm diameter pulley on the line shaft drives a 20 cm diameter pulley keyed to a dynamo shaft. Find the speed of the dynamo shaft when a. there is no slip .b. there is slip of 2.5% at each drive.[**Ans:** 2000 and 1901.25 rpm]

3. An engine, running at 150 r.p.m., drives a line shaft by means of a belt. The engine pulley is 750 mm diameter and the pulley on the line shaft being 450 mm. A 900 mm diameter pulley on the line shaft drives a 150 mm diameter pulley keyed to a dynamo shaft. Find the speed of the dynamo shaft, when 1. there is no slip, and 2. there is a slip of 2% at each drive. [**Ans:** 1500 rpm and 1440 rpm]

4. An engine shaft running at 120 r.p.m. is required to drive a machine shaft by means of a belt. The pulley on the engine shaft is of 2 m diameter and that of the machine shaft is 1 m diameter. If the

belt thickness is 5 mm ; determine the speed of the machine shaft, when 1. there is no slip ; and 2. there is a slip of 3%. [**Ans:** 239.4 r.p.m. ; 232.3 r.p.m.]

5. An engine shaft running at 120 r.p.m. is required to drive a machine shaft by means of a belt. The pulley on the engine shaft is of 2 m diameter and that of the machine shaft is 1 m diameter. If the belt thickness is 5 mm ; determine the speed of the machine shaft, when 1. there is no slip ; and 2. there is a slip of 3%. [**Ans:** 239.4 r.p.m. ; 232.3 r.p.m.]

Gear drives

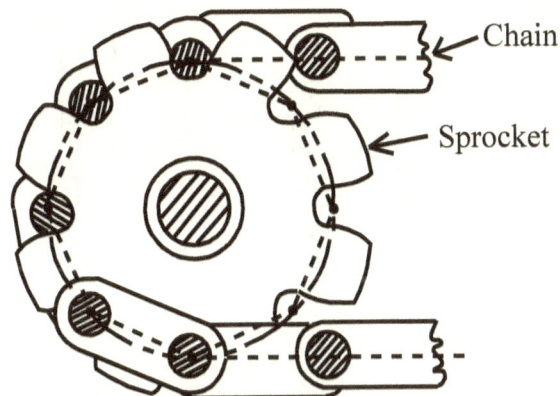

Fig. 10.11: Chain and sprockets

In order to avoid slipping, steel chains are used. The chains are made up of rigid links which are hinged together in order to provide the necessary flexibility for warping around the driving and driven wheels. The wheels have projecting teeth and fit into the corresponding recesses, in the links of the chain as shown in Fig. 10.11. The wheels and the chain are thus constrained to move together without slipping and ensures perfect velocity ratio. The toothed wheels are known as *sprocket wheels* or simply *sprockets.*

The chains are mostly used to transmit motion and power from one shaft to another, when the distance between the centres of the shafts is short such as in bicycles, motor cycles, agricultural machinery, road rollers, etc.

Terms used in gears

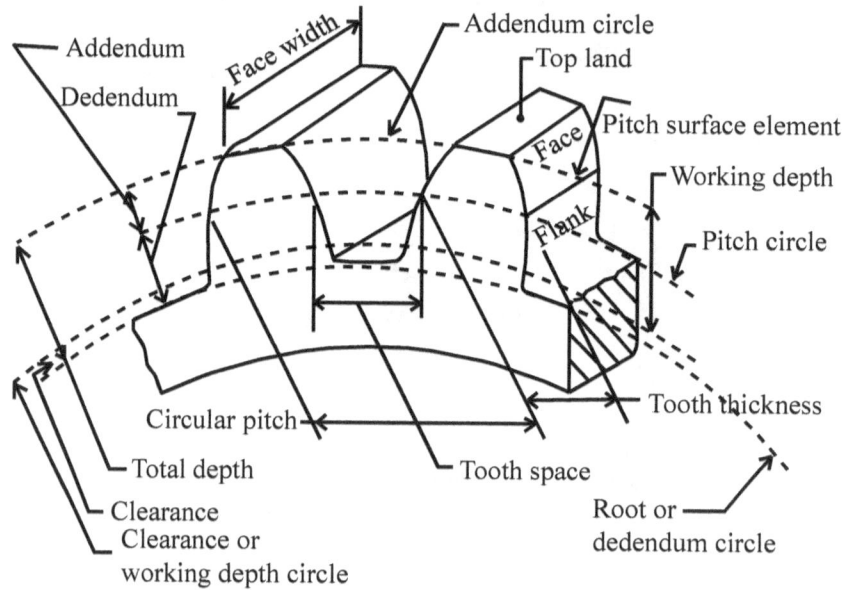

Fig. 10.12: Common terms used in gears

i. Pitch circle diameter

It is the diameter of the pitch circle.

ii. Addendum

It is a radial distance of a tooth from the pitch circle to the top of the teeth.

142

iii. Dedendum

It is the radial distance of a tooth from the pitch circle to the bottom of the tooth.

iv. Circular pitch

It is the distance measured on the circumference of the pitch circle from a point of one tooth to the corresponding point on the next tooth and usually denoted by P_c.

$$P_c = \frac{\pi D}{T} \quad \text{.........} \quad (10.13)$$

Where, D = Diameter of the pitch circle, and

       T = Number of teeth on the wheel.

v. Module

It is the ratio of the pitch circle diameter to the number of teeth and denoted by 'm'.

$$m = \frac{D}{T} \quad \text{.........} \quad (10.14)$$

Merits and demerits of gear over belt drives

Merits

i. As no slip takes place during chain drive, hence perfect velocity ratio is obtained.

ii. Since the chains are made of metal, therefore they occupy less space in width than a belt or rope drive.

iii. The chain drives may be used when the distance between the shafts is less.

iv. The chain drive gives a high transmission efficiency (upto 98 per cent).

v. The chain drive gives less load on the shafts.

vi. The chain drive has the ability of transmitting motion to several shafts by one chain only.

Demerits

i. The production cost of chains is relatively high.

ii. The chain drive needs accurate mounting and careful maintenance.

iii. The chain drive has velocity fluctuations especially when unduly stretched.

Bearing

A bearing is a machine element that support a moving part such as shafts and axles and confining to the desired motion.

Bearings are classified broadly according to the type of operation, the motions allowed, or to the directions of the loads applied to the parts.

Classification of bearings

Classification according to the principles of operation:

i. *Plain bearing*, known by the specific styles: bushing, journal bearing, sleeve bearing, etc.

ii. *Rolling- element bearing*, such as ball bearings and roller bearings.

iii. *Jewel bearing*, in which the load is carried by rolling the axle slightly off- centre.

iv. *Fluid bearing*, in which the load is carried by a gas or liquid.

v. *Magnetic bearing*, in which the load is carried by the magnetic field.

vi. *Flexure bearing*, in which the motion is supported by a load element which bends.

144

Materials used in bearings

i. Tin base babbits [Tin, Cu, Antimony, Pb]

ii. Lead base babbits [Pb, Sn, Antimony, Cu]

iii. Gun metal [Cu, Sn, Zn]

iv. Phosphorus bronze [Cu, Pb, Sn, P]

Properties of bearing materials

i. Should have high compressive strength to withstand maximum pressure.

ii. Should have high fatigue strength to withstand repeated loads.

iii. Should not have excessive wearing and heating.

iv. Should be able to accommodate position and work without being scoured by small particles of dust, grist, etc.

v. Should not corrode away under the action of lubricating oil.

vi. Should have high thermal conductivity for rapid removal of excess heat.

vii. Should have low coefficient of thermal expansion.

Ball and roller bearings

A ball and roller bearing is a type of rolling-element bearing that uses balls to maintain the separation between the bearing races.

The purpose of a ball and roller bearing is to reduce rotational friction and support radial and axial loads. It achieves this by using at least two races to contain the balls and transmit the loads through the balls. In most applications, one race is stationary and the other is attached to the rotating assembly (e.g., a hub or shaft). As one of the bearing races rotates it causes the balls to rotate as well. Because the balls are rolling they have a much lower coefficient of friction than if two flat surfaces were sliding against each other.

Ball and roller bearings tend to have lower load capacity for their size than other kinds of rolling-element bearings due to the smaller contact area between the balls and races. However, they can tolerate some misalignment of the inner and outer races.

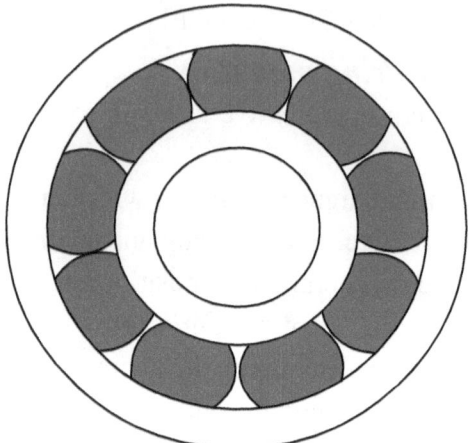

Fig. 10.13: Ball and roller bearing

The balls are generally made up of carbon chromium steel. The material of both the ball and races are heat treated to give extra hardness and toughness. The ball and rollers are manufactured by hot forging on hammer from steel rods. Then they are heat treated, ground and polished.

The lubricants are used in bearings to reduce friction between the rubbing surfaces, to carry away the heat generated by friction and to protect the bearings against corrosion.

Coupling

A coupling is a device used to connect two shafts together at their ends for transmitting power.

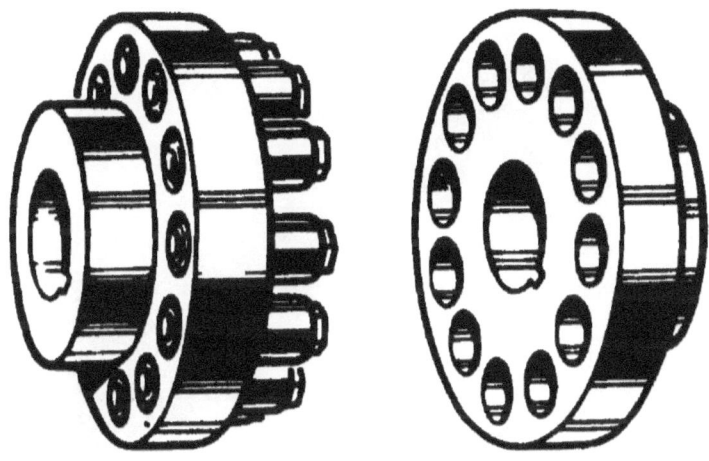

Fig. 10.14: Couplings

Uses

i. To connect driving and the driven part.

ii. To provide for the connection of shafts of units that are manufactured separately such as motor and generator and to provide for disconnection for repairs or alterations.

iii. To provide for misalignment of the shafts or to introduce mechanical flexibility.

iv. To reduce the transmission of shock loads from one shaft to another.

v. To protect against overloads.

vi. To alter the vibration characteristics of rotating units.

There are 2 types of couplings: rigid and flexible couplings. Rigid couplings are used when the shafts are parallel aligned whereas flexible couplings are aligned either lateral or angular.

Shafts

Fig. 10.15: Gear shaft

A shaft is a rotating machine element which is usually of circular cross- section used to transmit power from one place to another. The various members such as pulleys and gears are mounted on it.

Shafts are used for power transmission in engines, automobiles, machines, etc. For transmitting power from driving shafts to driven shafts, couplings, pulleys, gears, chain and sprocket, wheels, etc are used depending on the position of driving and driven shafts.

A good material for shaft and axle must have high strength, good machinability, and low sensitivity to concentration, withstand heat treatment operation and high wear resistance. The material used for ordinary shafts is mild steel. When high strength is required, an alloy steel such as nickel, nickel-chromium or chromium-vanadium steel is used.

Shafts are generally formed by hot rolling and finished to size by cold drawing or turning and grinding.

148

Crankshaft

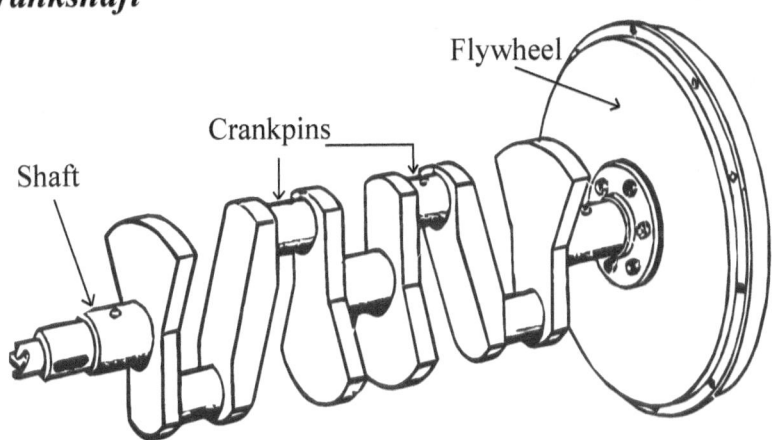

Fig. 10.16: Crankshaft

A crankshaft is the part of internal combustion engines that translates reciprocating linear piston motion into rotation. To convert the reciprocating motion into rotation, the crankshaft has 'crankpins', additional bearing surfaces whose aim is offset from that of the crank to which the big ends of the connecting rods from each cylinder attach. It is generally made much heavier and stronger necessary from the strength point of view so as to make requirements of rigidity and vibrations.

Large engines are usually multi cylinder to reduce pulsations from individual firing strokes, with more than one piston attached to a complex crankshaft. Many small engines, such as those found in mopeds or garden machinery, are single cylinder and use only a single piston, simplifying crankshaft design.

A crankshaft is subjected to enormous stresses, potentially equivalent of several tonnes of force. The crankshaft is connected to the fly-wheel (used to smooth out shock and convert energy to torque), the engine block, using bearings on the main journals, and to the pistons via their respective con-rods. An engine loses up to 75% of its generated energy in the form of friction, noise and vibration in the crankcase and piston area. The remaining losses occur in the valve train (timing chains, belts, pulleys, camshafts, lobes, valves, seals etc.) heat and blow by.

In industrial engines, 0.35% carbon steel and 0.45% chromium steel are commonly used, the rest is iron. In transport engines, magnetic steels is generally used. Aero- engines uses cranks made up of nickel- chromium steel.

Index

Evaporator, 14
Excitor, 126
Expansion Valve, 14

F

Fahrenheit scale, 71, 72
Fast and loose pulley drive, 134
Feed check valve, 35
Fire tube boiler, 26
Flexure bearing, 144
Fluid bearing, 144
fuel, 25, 26, 29, 31, 32, 33, 37,
   38, 39, 40, 41, 107
Fundamental or basic units, 2
fundamental quantities, 1, 2, 3, 4
Fuse, 126
Fusible plug, 34

G

Gate valve, 100
Gauge pressure, 81
Gear drives, 141
Gear pump, 112
generator, 125, 126, 147
Globe valve, 100, 101

H

HCV, 14, 40, 41
Heating and Humidifying, 59
Humidity ratio, 49, 50, 63

I

impeller, 107, 109, 110, 111, 112
Inclined tube manometer, 88

152

Inverted differential manometer,
   84, 94

J

Jewel bearing, 144

L

Lancashire boiler, 27, 28
Latent heat of fusion, 11
Latent heat of vaporization, 11
LCV, 40, 41

M

Magnetic bearing, 144
Mechanical power transmission,
   129, 163
Micro manometer, 84, 88, 90
Module, 143
Moist air, 43, 59
Multi- turn valve, 99
multistage centrifugal pump, 110

O

Open belt drive, 130, 131, 133
Optical pyrometer, 76, 77

P

physical quantity, 1, 2, 4
Piezometer tube, 82, 83
Pitch circle diameter, 142
Plain bearing, 144
positive displacement pumps, 108
Power factor, 122
Power transmitted by a belt, 137

Appendices

Appendix 1: Unit and conversion factors

Length	1 inch	= 0.0254 m
	1 ft	= 0.3048 m
Area	1 ft^2	= 0.0929m^2
Volume	1 ft^3	= 0.0283 m^3
	1 gal Imp	= 0.004546 m^3
	1 gal US	= 0.003785 m^3 = 3.785 litres
	1 litre	= 0.001 m^3
Mass	1 lb	= 0.4536 kg
	1 mole	molecular weight in kg
Density	1 lb/ft^3	= 16.03 kg m^{-3}
Velocity	1 ft/sec	= 0.3048 m s^{-1}
Pressure	1 lb/m^2	= 6894 Pa
	1 torr	= 133.3 Pa
	1 atm	= 1.013 x 10^5 Pa = 760 mm Hg
	1 Pa	= 1 N m^{-2} = 1 kg m^{-1} s^{-2}
Force	1 Newton	= 1 kg m s^{-2}
	1 lb ft s^{-2}	= 1.49 kg m s^{-2}
Viscosity	1 cP	= 0.001 N s m^{-2} = 0.001 Pa s
	1 lb/ft sec	= 1.49 N s m^{-2} = 1.49 kg m^{-1} s^{-2}
Energy	1 Btu	= 1055 J
	1 cal	= 4.186 J
Power	1 kW	= 1 kJ s-1
	1 W	= 1 J s-1
	1 horsepower	= 745.7 W = 745.7 J s^{-1} = 0.746 kW
	1 ton refrigeration	= 3.519 kW
Temperature units	(°F)	= 5/9 (°C) = 5/9 (K)
Heat-transfer coefficient	1 Btu ft^{-2} h^{-1} °F^{-1}	= 5.678 J m^{-2} s^{-1} °C
Thermal conductivity	1 Btu ft^{-1} h^{-1} °F^{-1}	= 1.731 J m^{-1} s^{-1} °C^{-1}
Constants	π	3.1416
	σ	5.73 x 10^{-8} J m^{-2}s^{-1}K^{-4}
	e (base of natural logs)	2.7183
	R	8.314 kJ mole^{-1} K^{-1} or 0.08206 m^3 atm mole^{-1} K^{-1}

(M) Mega = 10^6,
(k) kilo = 10^3,
(H) Hecto = 10^2
(m) milli = 10^{-3}
(μ) micro = 10^{-6}

Appendix 2: Pressure/enthalpy charts for refrigerants

(a) Tetrafluoroethane (refrigerant 134a)
reference state h = 200.0 kJ/kg; s = 1.00 kJ/(kg-K) for saturated liquid at 0°C

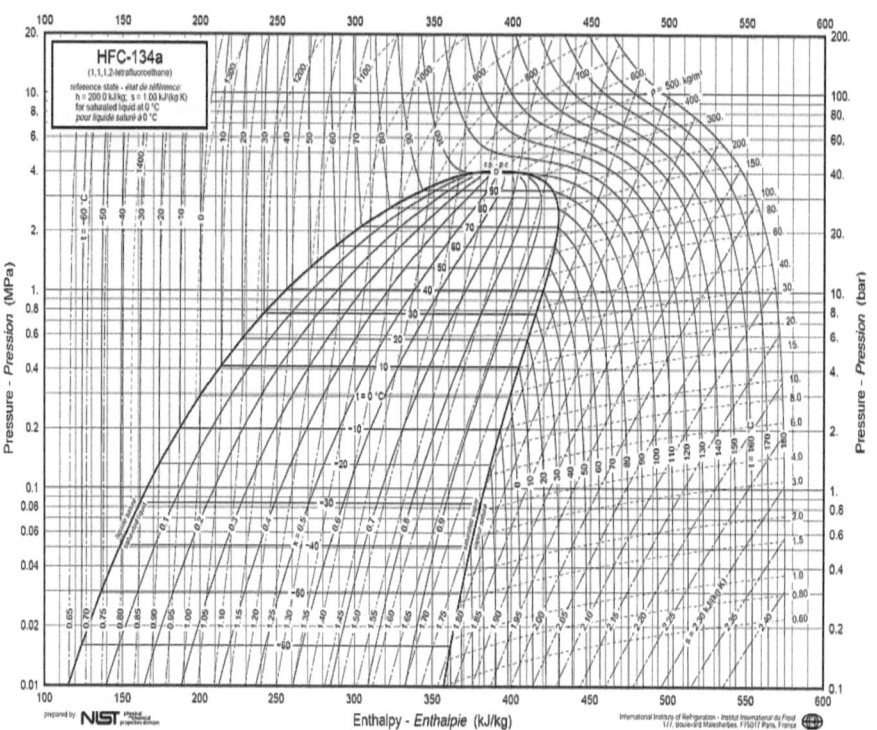

Enthalpy - *Enthalpie* (kJ/kg)

Source: The International Institute of Refrigeration (IIR) - *Institut International du Froid*, Paris, France. www.iifiir.org

156

(b) Ammonia (refrigerant 717)

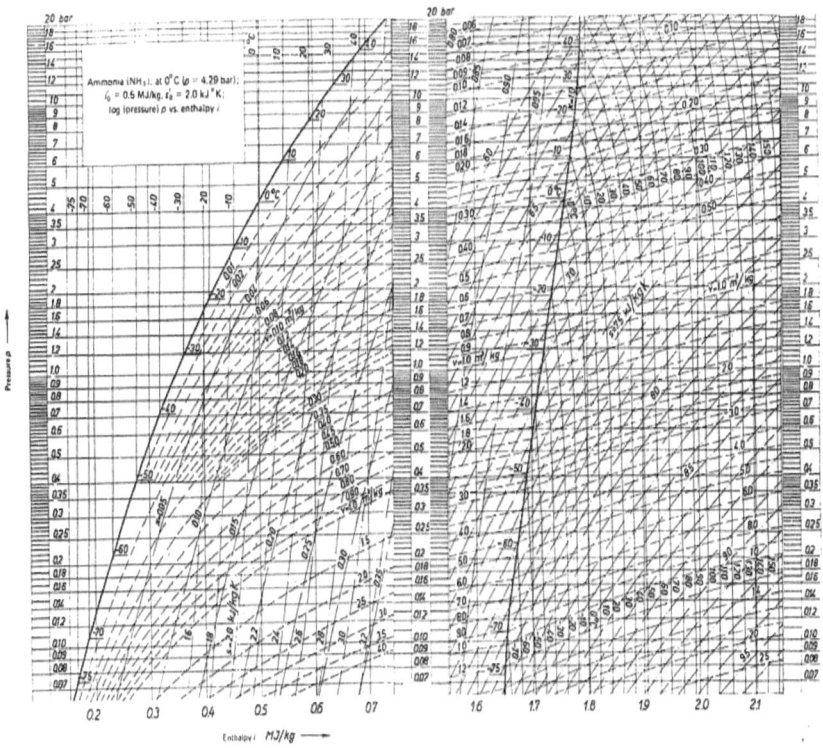

Source: *Kaltermachinen Regeln*, 5th edition, Verlag, C. F. Muller, Karlsruhe.

Appendix 3: Some properties of solids (Atm. Pressure)

	Thermal conductivity ($J\ m^{-1}\ s^{-1}\ °C^{-1}$)	Specific heat ($kJ\ kg^{-1}\ °C^{-1}$)	Density ($kg\ m^{-3}$)	Temperature ($°C$)
1. Metals				
Aluminium	220	0.87	2640	0
Brass	97	0.38	8650	0
Cast iron	55	0.42	7210	0
Copper	388	0.38	8900	0
Steel, mild	45	0.47	7840	18
Steel, stainless	21	0.48	7950	20
2. Non-metals				
Asbestos sheet	0.17	0.84	890	51
Brick	0.7	0.92	1760	20
Cardboard	0.07	1.26	640	20
Concrete	0.87	1.05	2000	20
Celluloid	0.21	1.55	1400	30
Cotton wool	0.04	1.26	80	30
Cork	0.043	1.55	160	30
Expanded rubber	0.04		72	0
Fibreboard insulation	0.052		240	21
Glass, soda	0.52	0.84	2240	20
Ice	2.25	2.10	920	0
Mineral wool	0.04		145	30
Polyethylene	0.55	2.30	950	20
Polystyrene foam	0.036		24	0
Polyurethane foam	0.026		32	0
Polyvinyl chloride	0.29	1.30	1400	20
Wood shavings	0.09	2.5	1.50	0
Wood	0.28	2.5	700	30

158

Appendix 4: Some properties of gases (Atm. Pressure)

	Thermal conductivity (J m^{-1} s^{-1} °C^{-1})	Specific heat (kJ kg^{-1} °C^{-1})	Density (kg m^{-3})	Temperature (°C)
Ammonia	0.022	2.19	0.73	15
Carbon dioxide	0.015	0.80	1.98	0
Refrigerant 134a (tetrafluoroethane)		1.46	1.21	25
Ammonia	0.022	2.19	0.73	15
Nitrogen	0.024	1.005	1.3	0

Appendix 5: Some properties of liquids (Atm. Pressure)

	Thermal conductivity (J m^{-1} s^{-1} °C^{-1})	Specific heat (kJ kg^{-1} °C^{-1})	Density (kg m^{-3})	Viscosity (N s m^{-2})	Temperature (°C)
Water (see Appendix 6)					0
Sucrose 20% soln.	0.54	3.8	1070	1.92×10^{-3}	20
				0.59×10^{-3}	80
60% soln.				6.2×10^{-3}	20
				5.4×10^{-3}	80
				3.7×10^{-3}	20
Sodium chloride 22% soln.	0.54	3.4	1240	2.7×10^{-3}	20
Acetic acid	0.17	2.2	1050	1.2×10^{-3}	20
Ethyl alcohol	0.18	2.3	790	1.2×10^{-3}	20
Glycerine	0.28	2.4	1250	830×10^{-3}	20
Olive oil	0.17	2.0	910	84×10^{-3}	20
Rape-seed oil			900	118×10^{-3}	20
Soya-bean oil			910	40×10^{-3}	30
Tallow			900	18×10^{-3}	65
Milk (whole)	0.56	3.9	1030	2.2×10^{-3}	20
Milk (skim)			1040	1.4×10^{-3}	25
Cream 20% fat			1010	6.2×10^{-3}	3
30% fat			1000	$13,8 \times 10^{-3}$	3

Appendix 6: Psychrometric chart (Low temperatures)

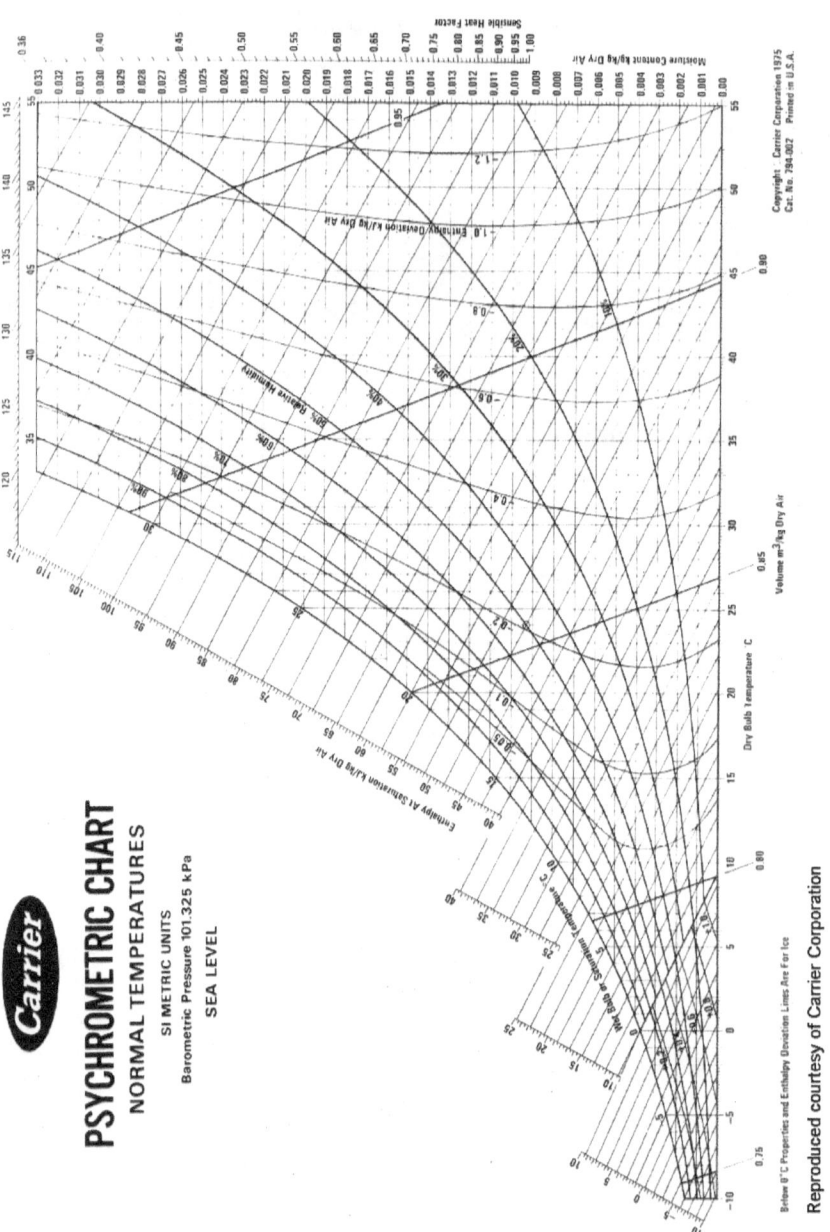

Reproduced courtesy of Carrier Corporation

160

Appendix 7: Psychrometric chart (High temperatures)

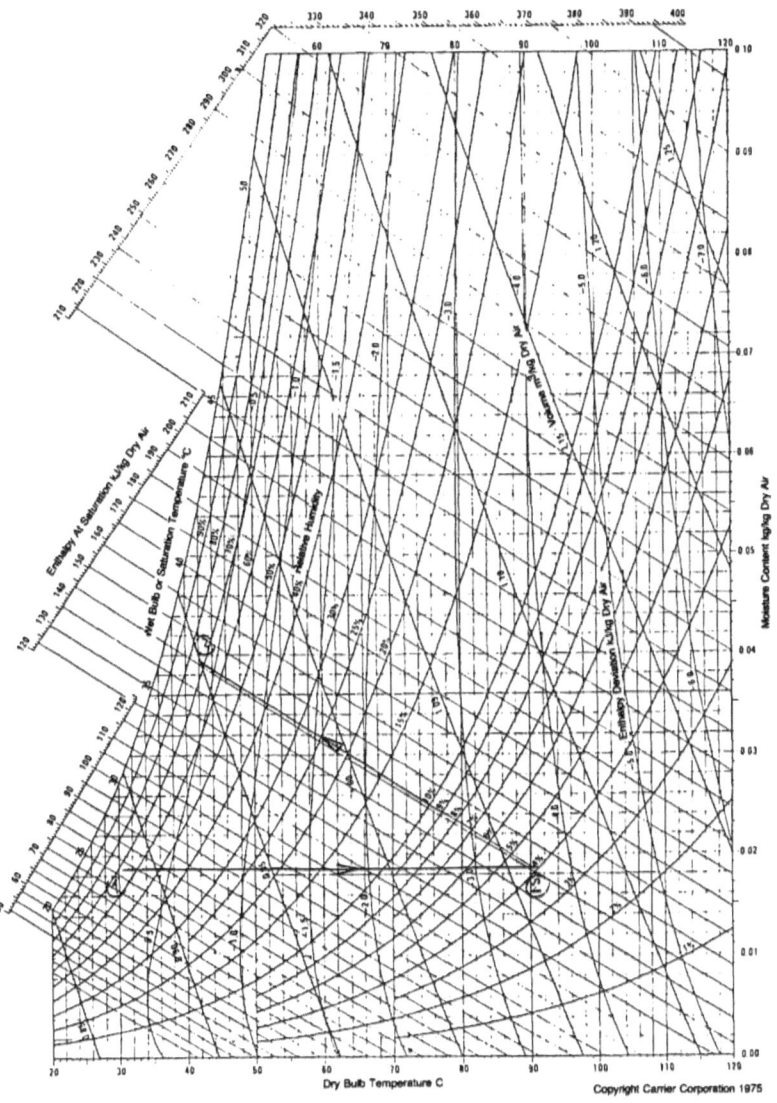

Dry Bulb Temperature C

Copyright Carrier Corporation 1975

161

Bibliography

1. Arora, R. C. (2010). Refrigeration and air conditioning: PHI Learning Pvt. Ltd.
2. Association, A. C. (2005). Standard handbook of chains: chains for power transmission and material handling: CRC Press.
3. Association, A. W. W. (1987). Basics of Electricity. Opflow, 13(2), 6-7.
4. Baldwin, A. J., & Hess, K. M. (1978). A programmed review of engineering fundamentals: Van Nostrand Reinhold Company.
5. Bansal, R. (2007). Engineering Mechanics: Laxmi Publications.
6. Brennen, C. E. (2011). Hydrodynamics of pumps: Cambridge University Press.
7. Brumbaugh, J. E. (2004). Audel HVAC Fundamentals: Volume 1: Heating Systems, Furnaces and Boilers (Vol. 17): John Wiley & Sons.
8. BSc, P. C. B. (1971). Mechanical power transmission
9. (1st ed.): The Macmillan Press Limited.
10. Dukelow, S. G. (1985). Improving boiler efficiency.
11. Earle, R. L. (2013). Unit operations in food processing: Elsevier.
12. Farrell, M., & Race, G. L. (2012). Practical psychrometry.CIBSE.
13. Handbook, A. (2001). Fundamentals. American Society of Heating, Refrigerating and Air Conditioning Engineers, Atlanta, 111.
14. Hundy, G. F., Trott, A. R., & Welch, T. (2008). Refrigeration and Air-conditioning: Butterworth-Heinemann.

15. Ipsen, D. C. (1960). Units, dimensions, and dimensionless numbers: McGraw-Hill.

16. Khurmi, R., & Gupta, J. (1976). Theory of machines: Eurasia.

17. Kuehn, K. (2014). A Student's Guide Through the Great Physics Texts: Volume II: Space, Time and Motion: Springer.

18. Lev, N. (1999). Centrifugal and Rotary Pumps: Fundamentals With Applications: CRC Press.

19. Miller, R., & Miller, M. R. (2006). Air conditioning and refrigeration: McGraw-Hill.

20. Moyers, C. G., & Baldwin, G. (1997). Psychrometry, evapourative cooling, and solids drying. Perry's chemical engineers' handbook, 1-90.

21. Nesbitt, B. (2011). Handbook of valves and actuators: valves manual international: Butterworth-Heinemann.

22. Papantonis, D. (2012). Centrifugal Pumps (1st ed.): InTech.

23. Radi, H. A., & Rasmussen, J. O. (2012). Principles of physics: for scientists and engineers: Springer Science & Business Media.

24. Rayaprolu, K. (2009). Boilers for power and process: CRC Press.

25. Refrigeration, & Heating, A.-C. E. A. S. o. (2010). 2010 ASHRAE Handbook: Refrigeration: American Society of Heating, Refrigerating and Air-conditioning Engineers, Incorporated.

26. Reynolds, P. (2013). Modern Power Station Practice: Incorporating Modern Power System Practice (Vol. 1000): Elsevier.

27. Ronney, P. D. (2006). Basics of Mechnical Engineering: Integrating Science , Technology and common sense.
28. Tullis, J. P. (1989). Hydraulics of pipelines: Pumps, valves, cavitation, transients: John Wiley & Sons.
29. Yanniotis, S. (2007). Solving problems in food engineering: Springer Science & Business Media.